Lecture Notes in Artifi

Subseries of Lecture Notes in Co
Edited by J. G. Carbonell and J. S.

'36

Lecture Notes in Computer Science

Edited by G. Goos, J. Hartmanis and J. van Leeuwen

Springer

Berlin
Heidelberg
New York
Barcelona
Budapest
Hong Kong
London
Milan
Paris
Santa Clara
Singapore
Tokyo

Elisabeth Maier   Marion Mast
Susann LuperFoy (Eds.)

Dialogue Processing in Spoken Language Systems

ECAI'96 Workshop
Budapest, Hungary, August 13, 1996
Revised Papers

 Springer

Series Editors

Jaime G. Carbonell, Carnegie Mellon University, Pittsburgh, PA, USA
Jörg Siekmann, University of Saarland, Saarbrücken, Germany

Volume Editors

Elisabeth Maier
DFKI GmbH
Stuhlsatzenhausweg 3, D-66123 Saarbrücken, Germany
E-mail: maier@dfki.uni-sb.de

Marion Mast
IBM Germany, European Speech Technology Group
Vangerowstraße 18, D-69115 Heidelberg, Germany
E-mail: mast@heidelbg.ibm.com

Susann LuperFoy
The MITRE Corporation
1820 Dolley Madison Boulevard, McLean, VA 22102, USA
E-mail: luperfoy@mitre.org

Cataloging-in-Publication Data applied for

Die Deutsche Bibliothek - CIP-Einheitsaufnahme

Dialogue processing in spoken language systems : revised papers /
ECAI '96 workshop, Budapest, Hungary, August 13, 1996. Elisabeth
Maier ... (ed.). - Berlin ; Heidelberg ; New York ; Barcelona ;
Budapest ; Hong Kong ; London ; Milan ; Paris ; Santa Clara ;
Singapore ; Tokyo : Springer, 1997
  (Lecture notes in computer science ; Vol. 1236 : Lecture notes in
  artificial intelligence)
  ISBN 3-540-63175-5

CR Subject Classification (1991): I.2.7

ISBN 3-540-63175-5 Springer-Verlag Berlin Heidelberg New York

© Springer-Verlag Berlin Heidelberg 1997
Printed in Germany

Typesetting: Camera ready by author
SPIN 10551817   06/3142 – 5 4 3 2 1 0   Printed on acid-free paper

Foreword

This volume contains a selection of extended and revised versions of papers presented to the European Conference on Artificial Intelligence ECAI-96 Workshop on Dialogue Processing in Spoken Language Systems. The workshop took place on August 13, 1996 in Budapest, Hungary. This workshop received considerable interest from researchers of internationally acclaimed research institutions and companies, indicating that spoken dialogue system development is a burning issue in the fields of artificial intelligence and speech processing.

This volume collects papers from researchers that belong to the leading groups in the community. It covers research being carried out in the United States (7 papers), in Europe (6 papers), and in Japan (1 paper). It is interesting to observe that international cooperation already exists in this field, as testified by one paper in the book which is authored jointly by researchers from Europe, the United States, and Japan. The book reports on work being pursued both in academia and in industry. The papers, though, are mostly biased towards application-oriented research, since most of the projects described in the book are either pursued in industry or heavily (co-)sponsored by companies interested in applications of speech processing technology.

This fact indicates that speech technology, and dialogue processing in particular, is a field that is in the process of transition from the research labs and is entering the marketplace: automatic speech-operated reservation and booking systems are already accessible to the public, as are information systems answering questions concerning timetables, weather conditions, private bank accounts, etc. Also, speech technology is more and more integrated into hands-free applications where a user is unable to employ interaction modalities like text or gestural interaction with graphic entities: typical applications are the operation of speech-controlled devices such as a communication radio or telephone in a car.

Along with the rapid advance of research in the area of spoken dialogue systems, progress is being made in many related fields: areas are identified where results from research in typed dialogues can be transferred to spoken language systems. With growing experience in system development, tasks like the collection and annotation of speech corpora are being made more efficient and less expensive. Wizard of Oz techniques are widely used both to involve the user in the system design from the earliest stages of development and to gather maximally realistic and usable data. Standards currently being developed for the transcription and annotation of speech (e.g., ToBi for phonetic labeling and the Discourse Resource Initiative (DRI) for segmentation and dialogue act labeling) lead to the reusability of speech corpora. According to the slogan "No data is better than more data" such massive data collections are crucial for system development, especially for training and testing of statistical models.

The treatment and recognition of prosodic features is of growing importance for spoken dialogue systems: since the segmentation of dialogue contributions into meaningful (processing) units is an important issue in spoken dialogue processing, the determination of such units, possibly by means of prosodic information like pauses and/or rising and falling intonation, is a topic which at the moment is receiving considerable attention.

As far as rapid prototyping is concerned, new design principles and architectures have been proposed that allow easy adaptation of dialogue systems to new domains and applications. For the most part, these requirements are met through the introduction of object-oriented design and by a clear-cut separation of application-dependent and application-independent knowledge.

Finally, the evaluation of spoken dialogue systems is a current topic that has been neglected in the past: while many studies have been carried out to assess the quality of speech recognition components, few efforts have been reported to estimate the overall quality of integrated spoken dialogue systems.

The book contains papers that address all of the problems mentioned above and propose methods for their solution.

The most challenging problem in speech processing is the transition from cooperative continuous speech input to truly spontaneous conversation. On the dialogue level, the transition from human-machine to human-human interaction, and eventually to multiparty discourse, is a central issue of current research. This book contributes substantially to the progress towards the next generation of spoken dialogue systems.

April 1997   Wolfgang Wahlster
                                                         Chair of ECAI-96

Contents

Overview

Elisabeth Maier[1], Marion Mast[2] and Susann LuperFoy[3]

[1] DFKI GmbH, Stuhlsatzenhausweg 3, D-66123 Saarbrücken
e-mail: maier@dfki.uni-sb.de
[2] IBM Germany, European Speech Technology Group, Vangerowstr. 18
D-69115 Heidelberg
e-mail: mast@heidelbg.ibm.com
[3] The MITRE Corporation, 1820 Dolley Madison Blvd, McLean, VA 22102 USA
e-mail: luperfoy@mitre.org

In recent years considerable progress has been made in the areas of speech recognition, natural language interpretation/generation, and dialogue processing for conversational interfaces to information retrieval systems. The chapters in this volume address a range of issues that impinge on the design, implementation, and evaluation of spoken language dialogue (SLD) systems. In a spoken language dialogue system a user's spoken input creates acoustic data that proceeds through a sequence of transitions: from acoustic signal to a sequence of word hypotheses, or more often a word hypothesis graph, from word hypothesis to syntactic and/or semantic analysis, and from context-independent interpretation into a system-specific command or query to be executed by the backend application system. The growing maturity of speech processing technology has made it feasible to begin constructing spoken dialogue interfaces to fairly complex systems. The development of this new sort of interface system requires an adjustment in research focus:

- from typed text to spoken input language
- from read speech to spontaneous speech and
- from simple recognition of read or transcribed speech to the interpretation and responsive action based on understanding of input speech

These new areas of endeavor require novel language processing techniques and the involvement of an innovative collection of knowledge sources and processing components.

As we consider the information available in the spoken language data stream we require a prosodic processing component to extract from the speech signal crucial information on stress, intonation contours, location and duration of pauses, and other features that allow the system to fully understand the user's message (see section2).

An important problem in the design and development of SLD systems is that of evaluation. For some of the separate fields, there exist well known and widely used evaluation criteria. In order to assign a comparative or a benchmark judgement to an SLD system, it is not sufficient to assemble independent evaluations of component elements. Otherwise, we might measure the word-error rate of the speech recognizer, compare the output of a sentential parse against a prepared key, or simply count the number of correct answers or behaviors exhibited by the backend system. In fact, the evaluation of SLD systems is an evaluation of a human-machine interface in which human behavior is a factor, in which time to task completion is not always the best metric of success. Rather, the whole system has to be taken into account (see section 4).

An important aspect to be considered during system evaluation is the user's experience of the system, i.e., whether it helps them to accomplish their task and whether they find it intuitive to work with. The social implications of an SLD system have to be studied carefully, to improve our chances of building systems which people will want to use. Some observations with respect to usability are collected in [Krause, 1997]. The following questions have to be considered, in order to build an effective interactive system:

- Who will be the potential users of the system ?
- Is the system easy to learn or intuitive to operate ?
- Can the system be used by inexperienced users ?
- Do users feel comfortable interacting with the system ?
- Does it help users to perform the intended tasks ?

1   FOUNDATIONS OF SPOKEN DIALOGUE SYSTEM DESIGN

Traditionally, for the development of a dialogue system one of the following approaches is adopted:

1. **Exploitation of user and task requirements**
   The potential users of the interface are interviewed in order to determine the requirements the interface has to fulfill; using these requirements, the specifications for the task to be carried out using the interface are generated. The interface is then developed following these specifications.
   This approach is taken when the task is new insofar as it does not have a prior history of conventions in human-human interaction from which the system developer can derive desiderata for the human-machine system. This occurs when characteristics of the computer actually introduce new capabilities that are not available in human-human dialogue interaction, e.g., the ability to keep a complete running record of all that has been said in the spoken dialogue thus far is the sort of thing that computers do well and humans do not.
   Typical examples for an interface developed along these lines can be found in the area of natural language or multimedia front ends: [Paris et al., 1995] describe a support tool for the (multilingual) drafting of instructions which describe how to compile pension forms; the tool has been designed in close cooperation with the designers of the administrative organizations that are responsible for the distribution of the forms. The MERIT system [Thiel et al., 1992] includes a user-oriented interface for information retrieval from a project data base.

2. **Use of models for human-human interaction**

The user interface is modeled after tasks that have traditionally been carried out by means of human-human interaction. The interface is designed on the basis of protocols that have been collected through tape recorded sessions involving two humans carrying out the same task without the aid of a computer. Typical examples can be found in areas like information seeking or service giving (e.g. reservation and booking of flights, hotels, etc.). Among such systems are EVAR [Mast *et al.*, 1992], which provides information concerning the train timetable, ATIS for flight reservation dialogues [Zue *et al.*, 1992], a system currently being developed at the University of Delft which gives information about public transport [van Vark *et al.*, 1997], or VERBMOBIL [Alexandersson *et al.*, 1997], where a translation system for spoken appointment scheduling dialogues has been modeled after human-human dialogues in the same domain.

3. **Use of models derived from Wizard of Oz studies**

In order to develop dialogue models for humans interacting with computer systems that do not yet exist (i.e., the systems we are striving to build) it is common to construct a data collection setup called the "Wizard of Oz" scenario (WOZ) in which a human experimenter plays the role of the computer and interacts with the subject. The data that results differ from that from human-human dialogue data collection since in the WOZ study the experimenter is behaving as the computer is expected to respond, and not as a human. Although the collection of data using this technique is very time consuming the methodology has recently been adopted in many projects since it allows the involvement of the user in an early stage of interface development. It also allows the user to get an impression of the interface functionality before the actual development of the system. This is of particular importance when the system introduces a new functionality that so far has been unknown to the user. Dialogue systems derived from WOZ studies have been developed in the WAXHOLM system [Carlson *et al.*, 1995], in the LINLIN project [Dahlbäck *et al.*, 1992], in the Danish Dialogue project [Dybkjær *et al.*, 1995], and partially also in VERBMOBIL [Bade *et al.*, 1994].

In all three cases interface development has to go through cycles of design, implementation, evaluation and re-design.

From what has been said so far it becomes clear that the development of dialogue systems relies highly on empirical observations of dialogue situations. Unfortunately, data-oriented design of dialogue systems tends to be both expensive and time-consuming. Therefore, it is highly desirable to adopt results and to re-use techniques that have been produced in previous research. Dahlbäck [Dahlbäck, 1997], for example, shows that some of the characteristics observed in spoken dialogues can also be found in non-spoken, i.e. written or typed dialogues. This means that principles developed for the design of typed dialogue systems can also be exploited for the development of spoken language dialogue systems. By factoring out a number of linguistic and non-linguistic aspects to characterize dialogue behavior Dahlbäck lays the foundation for a taxonomy of possible dialogue types.

Empirical and sociological studies are important not only for the design and the development of dialogue systems. They are also important for an assessment of the system quality and of its acceptance by the user; Krause [Krause, 1997] shows how factors like speed, correctness and completeness influence a user's willingness to work with a given dialogue system. By means of hidden-operator simulations he also shows how the user's problem solution behavior is changed by the introduction of the spoken dialogue capability. These observations give rise to suggestions for improvements to the system.

Also by means of a partially implemented and partially simulated system the authors of [Bernsen *et al.*, 1997] discover the types of errors the users make in a spoken dialogue system. It has been observed that errors occur, for example, when the user ignores previous moves made by the system or when the user deviates from the scenario.

2 DIALOGUE UNITS AND PROSODIC ASPECTS OF SPOKEN DIALOGUE PROCESSING

At first glance, the segmentation of a discourse into dialogue units and the analysis of prosodic aspects of dialogue are two separate areas of scientific study. In studying the structure of dialogues, however, the interrelation between prosody and dialogue units has become apparent. Since the segmentation information conferred by punctuation in written language is not available in the study of spontaneous speech, prosodic cues for segmentation become indispensable in addressing such questions as:

- what are the basic units in a dialogue
- how can these units be extracted from the consecutive ongoing utterances

Initial investigations of corpora, hand labeled with dialogue acts, are showing a coincidence of dialogue act boundaries with prosodically marked word boundaries (cf. [Batliner *et al.*, 1996b]).

Most of the work done so far on basic dialogue units refers to something like speech or dialogue acts (cf. [Bunt, 1981]). But there is no corresponding, well defined unit in the syntactic or semantic structure for speech acts and/or dialogue acts, and furthermore, spoken utterances tend to include many syntactically and semantically ill-formed utterances rendering the speech act or dialogue act insufficient for the purpose of analyzing spontaneous speech. [Lavie *et al.*, 1997] and [Traum and Heeman, 1997] discuss possible definitions of the utterance unit as a construct and the question of how they may be used to produce dialogue segmentation. In [Seligman *et al.*, 1997] the relation between pause units – which are segments bounded by natural pauses – and utterances is investigated.

Dialogue processing can also benefit from taking prosody into account when the issue is the identification of disfluencies which are observed to be frequent in spontaneous speech. For read or dictated speech quite acceptable word recognition rates have been achieved, e.g.

- over 90 % out-of-the box accuracy for large vocabulary, discrete dictation, speaker-independent (cf. [Kunzmann, 1996]) or
- 78 % for spontaneous spoken input (cf. [Schüssler *et al.*, 1997]).

But for the transition from typed to spoken input it is not enough to pass the speech input through a word recognizer which delivers the word hypothesis chain with 5 % error rate. Typed input can be corrected before feeding it to the system, whereas for spoken input each (self-) correction which occurs has to be detected and handled by the system. This means typed input is usually better formulated and less defective than spoken input (even with a nearly perfect word recognizer). Phenomena typical for spoken language are discussed in [Tischer, 1997]: Tischer investigated the syntax of self-corrections in spoken dialogues.

There are other contrasts between spoken and written language. In written language, beneath the content of the text, local and global structural phenomena, grammatical mood, speaker intentions, are often expressed through written indicators, including punctuation, but also choice of grammatical construction.

Hirschberg (cf. [Hirschberg and Grosz, 1992]) found that discourse structure is associated with intonational variation, but also that discourse segment boundaries in spoken language do not always align with paragraph boundaries in written text. She further discovered that multiple intonational features can be employed by a single speaker to convey a given structure element, e.g. longer pauses mark major topic shifts.

The transition from read to spontaneous speech results in the following phenomena:

- Hirschberg (cf. [Hirschberg and Nakatani, 1993]) found that self corrections and repairs appear in 10 % of spontaneous utterances and that acoustic-prosodic repair cues can be used for repair identification, word fragment identification and repair correction.
- The recognition of irregular boundaries which mark agrammatical phenomena like hesitations, repairs etc. is described in [Kießling *et al.*, 1996, Batliner *et al.*, 1996b].
- The speaking rate can be highly inconsistent within utterances and across utterances, as well as within a session and across sessions and speakers. (cf. [Butzberger *et al.*, 1992]).
- The articulation can be highly variable with elicitations of many function words (e.g. unusual lengthening and stress patterns).

Of course this list is not complete, but it shows which progress was made to obtain systems which can communicate with a user via natural dialogue. On the other side it shows the remaining problems which have to be solved in order to succeed in developing systems with which the user feels comfortable.

Besides this *"... prosodic parameters can be used on all levels in the decoding process of an ASR* [automatic speech recognition] *system, for example, at the acoustic-phonetic level ... the lexical level ... the syntactic level ... and at the pragmatic level ..."* [Lea *et al.*, 1975].

Results from empirical study of the interaction between prosody and dialogue are already being successfully exploited in the following ways:

- Disambiguation with respect to morphology, the word level, syntax, semantics and pragmatics, e.g. a different word accent can change a noun to a verb (cf. [Nöth *et al.*, 1996, Kießling *et al.*, 1994, Kießling *et al.*, 1996, Batliner *et al.*, 1996a]).
- Determination of the sentence modality (cf. [Kießling *et al.*, 1993, Kompe *et al.*, 1994]).
- Increase in processing speed during syntactic parsing through the pruning of parse alternatives which are incompatible with the prosodic structure (cf. [Bakenecker *et al.*, 1994, Kompe *et al.*, 1997]).

Another field where prosody can improve SLD system performance is speech production. For example, [Grote *et al.*, 1997] presents an approach for the selection of appropriate intonation for speech production in dialogue systems.

3 SPOKEN LANGUAGE DIALOGUE SYSTEMS – DESIGN AND IMPLEMENTATION

As more and more systems enter the market issues like rapid prototyping and portability of spoken dialogue systems are of growing importance. While the individual methodologies and techniques for the development of spoken dialogue systems are well explored the integration of such spoken dialogue interfaces into new application domains and with new systems is often problematic.

Various approaches have been proposed to minimize the effort and to maximize the knowledge reuse when adopting a spoken dialogue system to a new domain; among the most prominent approaches are:

- **Separation of domain-dependent and domain-independent information**
  Human dialogue behavior has many commonalities even across many different domains and tasks; it is intrinsic to every dialogue that the interactants engage in an exchange of questions and responses, raise and satisfy expectations and carry out actions. Therefore, a core of domain-independent dialogue strategies can be identified and implemented for use in different applications. For every new application these strategies can be specified further in order to meet the requirements of the domain: every domain-dependent strategy can be defined in terms of less specific domain-independent principles to which domain-dependent behaviors are added.
- **Development of an object-oriented system architecture**
  An object-oriented architecture has proven to be appropriate for a system that has to be frequently adapted to new domains and applications: such a system disposes of a number of system components with a well-defined functionality. In principle, individual modules can be replaced by components that better serve the needs of each new application. The only requirement

to be met by the new modules is that they communicate with the overall system along well-defined interface protocols. Another advantage of an object oriented system design is that components can be tailored towards new domains by adding further specializations. Methods defined on an abstract level remain valid over the whole module. Only domain-specific extensions need to be added where necessary.

– **Reuse of commonly accepted categories and technologies**
  If similar applications exist both theoretical and practical solutions may be adapted and reused, thereby saving considerable amounts of development time. Categories and technologies may also be adapted from entirely different applications: typical categories that carry across different domains are common strategies for dialogue behavior; an example for technologies that can be used in a whole range of dialogue systems are e.g. automatic learning methods for dialogue models or techniques for the prediction of dialogue states [Reithinger and Maier, 1995].

All of the above design principles can only be fulfilled if all components are specified declaratively.

A system that has clearly separated domain-dependent from domain-independent information is described in [Barnett and Singh, 1997]: the system consists of four application-independent subcomponents: a domain model contains the world knowledge, a transaction component which includes information necessary to solve tasks, a speech recognition module for the treatment of user input, and a dialogue model that defines the legal moves of a dialogue. All these components only contain a limited amount of information which then has to be customized to the specific applications. All subcomponents are designed to ensure robust processing: the system incorporates methods for error recovery.

A system that develops similar design strategies is described in [Abella *et al.*, 1997]: here an object-oriented architecture for spoken dialogue processing is proposed. The system incorporates a number of dialogue principles that cover a broad range of potential applications. Among these principles are, for example, disambiguation of user input, error correction, relaxation of overspecified and augmentation of underspecified information. A dialogue manager includes objects to handle these principles and to apply them in a meaningful, efficient and inexpensive manner.

A design problem that is specific to *spoken* dialogue systems is addressed in [Qu *et al.*, 1997]: the authors show how problems related to speech processing, such as missing information and failed recognition, can lead to cumulative errors in dialogue processing. Two methods are proposed which combine contextual, grammatical and statistical information in order to minimize the influence of errors on the overall processing task.

4 EVALUATION OF SYSTEMS

Now that there are usable SLD systems in research labs but also accessible to the public in off-the-shelf products, the need to evaluate such systems is increasingly important.

Beside the comparative evaluation of competing SLD techniques and systems, an evaluation methodology should be able to measure

- which changes in the design or the components of the system improve a given system over time, and
- the benefits and advantages that a potential user will derive from using the system.

Hirschman and Thompson ([Hirschman and Thompson, 1994], p. 479) observe that: " ... , *evaluation plays an important role for system developers (to tell if their system is improving), for system integrators (to determine which approaches should be used where) and for consumers (to identify which system will best meet a specific set of needs). Beyond this, evaluation plays a critical role in guiding and focusing research.*"

In the evaluation section of the EAGLES-Report [Hirschman and Thompson, 1994] three kinds of evaluations are proposed:

- **adequacy evaluation**, which is the determination of the fitness of a system for a purpose. From it, potential users can benefit in being able to judge and compare systems.
  "... , it is becoming increasingly important to address the legitimate needs of potential users in determining whether any of the products on offer in a given application domain are adequate for their particular task, ..." ([Hirschman and Thompson, 1994], p. 476).
- **diagnostic evaluation**, which is the production of a system performance profile, mostly done with test suites of exemplary input.
  "Test suites are particularly valuable to system developers and maintainers, allowing automated regression testing to ensure that system changes have the intended effect and no others, ..." ([Hirschman and Thompson, 1994], p. 477).
- **performance evaluation**, which is the measurement of system performance in one or more specific areas.
  " It is typically used to compare like with like, whether two alternative implementations of a technology, or successive generations of the same implementation." ([Hirschman and Thompson, 1994], p. 476).

Given these competing objectives for SLD system evaluation, there are two distinct procedures for evaluation, **glass box** and **black box**.
"A distinction is often drawn between so-called glass box and black box evaluation, which sometimes appears to differentiate between component-wise versus whole-system evaluation, ... " [Hirschman and Thompson, 1994].

For some of the subtasks which have to be solved in an SLD system, well established evaluation methods can be employed. These include the simple word accuracy rate for the speech recognition and the percentage of correctly answered questions for a written language database retrieval system. Additional evaluation measures have been proposed and have undergone preliminary testing, but there is still no consensus on a single set of evaluation metrics for SLD systems in general.

A possible evaluation methodology for SLD systems would be to normalize and sum the votes for all the modules. This results in a glass box evaluation. But since the whole is more than the sum of its parts, this judgement must not be very trustworthy. E.g. a system with a rather simple semantic interpretation based on a few keywords can be of greater use than one with a more complete and more general semantic interpretation which can discern subtle nuances in meaning, in combination with a rather bad word recognizer or even a good recognizer that expects grammatical input when applied to (ungrammatical) spontaneous speech. For such a query system it may be more efficacious to extract the important keywords from the input instead of interpreting every word in order to produce a complete semantic analysis.

Another difficulty in comparing SLD systems is the case where the systems to be compared do not consist of the same configuration of modules. In this case, it is better to judge the overall performance of each system as a whole without testing the individual components. This is known as a black box evaluation. In [Gates et al., 1997] an end-to-end evaluation method is proposed to evaluate and compare different translation systems.

In this case however the evaluation criteria must be carefully specified in advance to avoid an apples-to-oranges comparison of competing systems that happen to have been designed to embody different strengths and emphasize efficacy for different tasks. The comparison of systems is not an easy thing to do even when all systems handle the same task, that is, the systems have to be at the same place in order to make them available for the same test team. Ideally each system should be developed using the same training data and tested against the same test data. First, a given input could legitimately be met with a number of different, yet equally valid dialogue responses. Second, the prior system output will affect the user's next spoken input, which will in turn condition the requirements for the next system output. Thus it often makes sense to use this measure for the first utterance in each dialogue only. After that the system reaction has to be taken into account for the subsequent dialogue contributions. A new automatic evaluation environment for SLD systems is proposed in [Araki and Doshita, 1997] where two systems are evaluated during communication with an automatic coordinator which creates a log file of the dialogue.

Qualities of a dialogue interface system that are normally summarised by the term 'user satisfaction' give indicators of the system's perceived usefulness and usability according to the intended user group. This includes, whether the user

- gets the information s/he wants
- is comfortable with the system
- gets the information within an acceptable elapsed time, etc.

Interesting issues are, e.g., which input and output modalities users prefer. Depending on the task, users may prefer either written or spoken language for input and/or output. In [Sikorski and Allen, 1997] authors present recent experimental results which explore the effects of word recognition accuracy and speech versus keyboard input on task performance.

A measure of user satisfaction can be obtained through a test-user questionnaire designed to elicit indicators of the sort of opinions listed above. This obviously results in a subjective judgement of the system.

5 Acknowledgements

We are grateful to the members of the program committee who helped us to put together an interesting workshop program: Nick Campbell, Morena Danieli, Norman Fraser, and Julia Hirschberg. We also would like to thank the participants of the ECAI-96 workshop on Dialogue Processing in Spoken Language Systems for inspiring discussions.

References

[Abella et al., 1997] Alicia Abella, Michael K. Brown, and Bruce Buntschuh. Development Principles for Dialog-Based Interfaces, 1997. *This volume.*

[Alexandersson et al., 1997] Jan Alexandersson, Norbert Reithinger, and Elisabeth Maier. Insights into the Dialogue Processing of VERBMOBIL. In *Proceedings of ANLP 1997*, Washington, 1997.

[Araki and Doshita, 1997] M. Araki and S. Doshita. Automatic Evaluation Environment for Spoken Dialogue Systems, 1997. *This volume.*

[Bade et al., 1994] Ute Bade, Susanne Heizmann, Susanne Jekat-Rommel, Shinichi Kameyama, Detlev Krause, Ilona Maleck, Birte Prahl, and Wiebke Preuss. Der Verbmobilsimulator und Wizard of Oz-Experimente in TP 13, Juni 1994. Verbmobil Memo No. 24, in German.

[Bakenecker et al., 1994] G. Bakenecker, U. Block, A. Batliner, R. Kompe, E. Nöth, and P. Regel-Brietzmann. Improving Parsing by Incorporating 'Prosodic Clause Boundaries' into a Grammar. In *Int. Conf. on Spoken Language Processing*, volume 3, pages 1115-1118, Yokohama, September 1994.

[Barnett and Singh, 1997] James Barnett and Mona Singh. Designing a portable spoken dialogue system, 1997. *This volume.*

[Batliner et al., 1996a] A. Batliner, A. Feldhaus, S. Geissler, A. Kießling, T. Kiss, R. Kompe, and E. Nöth. Integrating Syntactic and Prosodic Information for the Efficient Detection of Empty Categories. In *Proc. of the Int. Conf. on Computational Linguistics*, volume 1, pages 71-76, Kopenhagen, August 1996.

[Batliner et al., 1996b] A. Batliner, R. Kompe, A. Kießling, M. Mast, and E. Nöth. All about Ms and Is, not to forget As, and a comparison with Bs and Ss and Ds. Towards a syntactic–prosodic labeling system for large spontaneous speech data bases. Verbmobil Memo 102, Februar 1996.

[Bernsen et al., 1997] Niels Ole Bernsen, Laila Dybkjær, and Hans Dybkjær. User errors in spoken human-machine dialogue, 1997. *This volume.*

[Bunt, 1981] H. Bunt. Rules for the interpretation, evaluation and generation of dialogue acts. In *IPO Annual Progress Report 16*, pages 99 - 107, 1981.

[Butzberger et al., 1992] J. Butzberger, H. Murveit, E. Shriberg, and P. Price. Modeling Spontaneous Speech Effects in Large Vocabulary Speech Recognition Applications. In *Speech and Natural Language Workshop*. Morgan Kaufmann, 1992.

[Carlson et al., 1995] Rolf Carlson, Sheri Hunnicutt, and Joakim Gustafsson. Dialog Management in the WAXHOLM System. In *Proceedings of the ESCA Workshop on Spoken Dialogue Systems - Theories and Applications*, pages 137–140, Vigsø, Denmark, 1995.

[Dahlbäck et al., 1992] Nils Dahlbäck, Arne Jönsson, and Lars Ahrenberg. Wizard of Oz-studies - why and how. Research Report LiTH–IDA–R–92–19, Institutionen för Datavetenskap, Universitetet och Tekniska Högskolan Linköping, August 1992. Also published in Proceedings of the 14th Annual Conference of the Cognitive Sience Society (COG SCI–92) Bloomington, Indiana, July 29th to August 1st.

[Dahlbäck, 1997] Nils Dahlbäck. Towards a Dialogue Taxonomy, 1997. *This volume.*

[Dybkjær et al., 1995] Hans Dybkjær, Niels Ole Bernsen, Laila Dybkjær, and Dimitris Papazachariou. Wizard-of-Oz and the Trade-Off between Naturalness and Recogniser Constraints. In *Proceedings of EUROSPEECH '93*, pages 947–950, Berlin, September 1995.

[Gates et al., 1997] D. Gates, A. Lavie, L. Levin, A. Waibel, M. Gavalda, L. Mayfield, M. Woszczyna, and P. Zhan. End-to-end Evaluation in JANUS: a Speech-to-speech Translation System , 1997. *This volume.*

[Grote et al., 1997] B. Grote, E. Hagen, A. Stein, and E. Teich. Speech Production in Human-Machine Dialogue: A Natural Language Generation Perspective, 1997. *This volume.*

[Hirschberg and Grosz, 1992] J. Hirschberg and B. Grosz. Intonational Features of Local and Global Discourse Structure. In *Proc. Speech and Natural Language Workshop*, pages 441–446, Harriman NY, 1992. Morgan Kaufman.

[Hirschberg and Nakatani, 1993] J. Hirschberg and C. Nakatani. A Speech–First Model for Repair Identification in Spoken Language Systems. In *Proc. European Conf. on Speech Communication and Technology*, volume 2, pages 1173–1176, Berlin, September 1993.

[Hirschman and Thompson, 1994] L. Hirschman and H. Thompson. Evaluation. In R. Cole, J. Mariani, A. Uszkoreit, H. Zaenen, and V. Zue, editors, *Survey of the State of the Art in Human Language Technology*, pages 475–518. 1994.

[Kießling et al., 1993] A. Kießling, R. Kompe, H. Niemann, E. Nöth, and A. Batliner. "Roger", "Sorry", "I'm still listening": Dialog guiding signals in information retrieval dialogs . In D. House and P. Touati, editors, *Proc. of an ESCA Workshop on Prosody*, pages 140–143. Lund University, Department of Linguistics, Lund, September 1993.

[Kießling et al., 1994] A. Kießling, R. Kompe, H. Niemann, E. Nöth, and A. Batliner. Detection of Phrase Boundaries and Accents. In H. Niemann, R. De Mori, and G. Hanrieder, editors, *Progress and Prospects of Speech Research and Technology: Proc. of the CRIM / FORWISS Workshop*, PAI 1, pages 266–269. Infix, September 1994.

[Kießling et al., 1996] A. Kießling, R. Kompe, A. Batliner, H. Niemann, and E. Nöth. Classification of Boundaries and Accents in Spontaneous Speech. In R. Kuhn, editor, *Proc. of the 3rd CRIM / FORWISS Workshop*, pages 104–113, Montreal, October 1996.

[Kompe et al., 1994] R. Kompe, E. Nöth, A. Kießling, T. Kuhn, M. Mast, H. Niemann, K. Ott, and A. Batliner. Prosody takes over: Towards a prosodically guided dialog system. *Speech Communication*, 15(1–2):155–167, Oktober 1994.

[Kompe et al., 1997] R. Kompe, A. Kießling, H. Niemann, E. Nöth, A. Batliner, S. Schachtl, T. Ruland, and H.U. Block. Improving Parsing of Spontaneous Speech with the Help of Prosodic Boundaries. In *Proc. Int. Conf. on Acoustics, Speech and Signal Processing*, München, 1997. (to appear).

[Krause, 1997] Detlev Krause. Using an Interpretation System - Some Observations in Hidden Operator Simulations of VERBMOBIL, 1997. *This volume.*

[Kunzmann, 1996] S. Kunzmann. The VoiceType family: A multilingual speech recognition system. In *Proceedings of the 3rd CRIM-FORWISS Workshop, Montreal, Canada (7.- 8. October)*, 1996.

[Lavie et al., 1997] A. Lavie, D. Gates, N. Coccaro, and L. Levin. Input Segmentation of Spontaneous Speech in JANUS: a Speech-to-speech Translation System , 1997. *This volume.*

[Lea et al., 1975] W.A. Lea, M.F. Medress, and T.E. Skinner. A prosodically guided speech understanding strategy. *IEEE Trans.*, ASSP-23:30–38, 1975.

[Mast et al., 1992] M. Mast, R. Kompe, F. Kummert, H. Niemann, and E. Nöth. The dialog module of the speech recognition and dialog system EVAR. In *Proceedings of the International Conference on Spoken Language Processing, Banff, Canada*, pages 1573–1576, 1992.

[Nöth et al., 1996] E. Nöth, R. Kompe, A. Kießling, H. Niemann, A. Batliner, S. Schachtl, T. Ruland, and U. Block. Prosodic Parsing of Spontaneous Speech . In J.-P. Haton, editor, *Proc. of the Workshop on Multi-lingual Spontaneous Speech Recognition in Real Environments*, Nancy, June 1996. CRIN/CNRS–INRIA.

[Paris et al., 1995] Cécile Paris, Keith Vander Linden, Markus Fischer, Anthony Hartley, Lyn Pemberton, Richard Power, and Donia Scott. A support tool for writing multilingual instructions. In *Proceedings of IJCAI-95*, Montreal, Canada, August 1995.

[Qu et al., 1997] Yan Qu, Barbara Di Eugenio, Alon Lavie, and Carolyn P. Rosé. Minimizing Cumulative Error in Discourse Context, 1997. *This volume.*

[Reithinger and Maier, 1995] Norbert Reithinger and Elisabeth Maier. Utilizing Statistical Speech Act Processing in VERBMOBIL. In *Proceedings of the 33rd Annual Meeting of the Association for Computational Linguistics (ACL-95)*, Cambridge, MA, June 1995.

[Schüssler et al., 1997] M. Schüssler, F. Gallwitz, and S. Harbeck. A Fast Algorithm for Unsupervised Incremental Speaker Adaptation. In *Proc. Int. Conf. on Acoustics, Speech and Signal Processing*, München, 1997. (to appear).

[Seligman et al., 1997] M. Seligman, J. Hosaka, and H. Singer. "Pause Units" and Analysis of Spontaneous Japanese Dialogues: Preliminary Studies, 1997. *This volume.*

[Sikorski and Allen, 1997] T. Sikorski and J. Allen. A Task-Based Evaluation of the TRAINS-95 Dialogue System, 1997. *This volume.*

[Thiel et al., 1992] Ulrich Thiel, Matthias Hemmje, Anja Kerner, Martin Kracker, Stefan Sitter, Adelheit Stein, and Anne Tißen. Towards a user-centered interface for information retrieval: The MERIT system. In *Proceedings of SIGIR*, 1992.

[Tischer, 1997] B. Tischer. Syntactic Procedures for the Detection of Self-Repairs in German Dialogues, 1997. *This volume.*

[Traum and Heeman, 1997] D. Traum and P. Heeman. Utterance Units in Spoken Dialogue, 1997. *This volume.*

[van Vark et al., 1997] R.J. van Vark, J.P.M. de Vreught, and L.J.M. Rothkranz. Classification of Public Transport Information Dialogues Using an Information Based Coding Scheme, 1997. *This volume.*

[Zue *et al.*, 1992] Victor Zue, James Glass, David Goodeau, Lynnette Hirschman, Michael Philips, Joseph Polifroni, and Stephanie Seneff. The MIT ATIS System. Technical report, Office of Naval Research, Harriman, NY, February 1992.

User Errors in Spoken Human-Machine Dialogue

Niels Ole Bernsen, Laila Dybkjær and Hans Dybkjær

Centre for Cognitive Science, Roskilde University,

PO Box 260, 4000 Roskilde, Denmark

phone: +45 46 75 77 11   fax: +45 46 75 45 02

emails: nob@cog.ruc.dk, laila@cog.ruc.dk, dybkjaer@cog.ruc.dk

Abstract

Controlled user testing of the dialogue component of spoken language dialogue systems (SLDSs) has a natural focus on the detection, analysis and repair of dialogue design problems. Not only dialogue designers and their systems commit errors, however. Users do so as well. Improvement of dialogue interaction is not only a matter of reducing the number and severity of dialogue design problems but also of preventing the occurrence of avoidable user errors. Based on a controlled user test of the dialogue component of an implemented SLDS, the paper takes a systematic look at the dialogue errors made by users in the test corpus. A typology of user errors in spoken human-machine dialogue is presented and discussed, and potentially important dialogue design advice derived from the fact that the notion of a 'user error' turns out to be one that must be handled with care.

1 Introduction

This paper is based on a controlled user test of the dialogue component of the Danish dialogue system which is an advanced spoken language dialogue system (SLDS). When analysing the data from such tests, the natural focus is on dialogue design errors. Such errors have to be identified and diagnosed, and ways of remedying them must be found whenever possible. Dialogue design errors cause problems of user-system interaction, make user task performance unnecessarily bumpy and generate user dissatisfaction with SLDS technology. However, not everything that goes wrong in the dialogue between user and system is due solely to errors made by the dialogue designers. As an extreme viewpoint one could of course maintain that a system must be able to deliver what the user wants no matter how the user behaves. However, we suspect that not even a mind-reading system could do that. At the other extreme one could claim that users just have to get used to the system no matter how silly the system behaves. But nobody would claim that, we submit. In other words, a more balanced viewpoint is required. Users also make errors during dialogue and some interaction problems are the compound effects of dialogue design errors and user errors. This paper proposes to take a systematic look at the dialogue errors made *solely* by users. As we shall see, it is not always straightforward to separate errors made solely by users from compound errors and from pure errors of dialogue design, even if one starts with what is apparently a reasonable definition of 'user error'. In this paper we want to present progress made with respect to the question of how to properly define a user error and distinguish such errors from design errors and compound design and user errors. Based on the user test material, we ask: of which types are the user errors that were identified? What are their likely effects on the success of the dialogue? What, if anything, can be done about them? And what is a 'user error in human-machine dialogue' in the first place?

The Danish SLDS prototype is a ticket reservation system for Danish domestic flights. The system runs on a PC with a DSP board and is accessed over the telephone. It is a walk-up-and-use application which understands speaker-independent continuous spoken Danish with a vocabulary of about 500 words. The prototype runs in close-to-real-time and is representative of advanced current systems. Comparable SLDSs are found in [1,3,7]. The system has five main modules. The *speech recogniser* produces a 1-best string of words. The *parser* makes a syntactic analysis of the string and extracts the semantic contents which are represented in frame-like structures. The *dialogue handling module* interprets the contents of the semantic objects and decides on the next system action which may be to send a query to the *application database*, send output to the user, or wait for new input. In the latter case, predictions on the next user input are sent to the recogniser and the parser. *Output* is produced by concatenating pre-recorded phrases under the control of the dialogue module.

In what follows, Section 2 provides a description of the dialogue model for the Danish dialogue system and presents an example dialogue from the user test. The user test is described in Section 3. Section 4 presents an analysis of the user errors that were identified in the user test. Section 5 concludes the paper.

2 The Dialogue Model

The dialogue model for the Danish dialogue system was developed by the Wizard of Oz (WOZ) experimental prototyping method in which a person simulates the system to be designed in dialogue with users who are made to believe that they interact with a real system [8]. The dialogue model had to satisfy the following technological constraints imposed by the speech recogniser: to ensure real-time performance, at most 100 words could be active in memory at a time; to ensure an acceptable recognition rate, an average and a maximum user utterance length of 3-4 words and 10 words, respectively, were imposed. Other design goals, such as linguistic naturalness, dialogue naturalness and dialogue flexibility had to be traded off against these constraints [4].

The WOZ dialogue model development was iterated until the model satisfied the design constraints. In each iteration, the dialogues were recorded, transcribed, analysed and used as a basis for improvements on the dialogue model. We performed seven WOZ iterations yielding a transcribed corpus of 125 task-oriented human-machine dialogues corresponding to approximately seven hours of spoken dialogue. The 94 dialogues that were recorded during the last two iterations were performed by external subjects whereas only system designers and colleagues had participated in the earlier iterations. A total of 24 different subjects were involved in the seven iterations. Dialogues were based on written descriptions of reservation tasks (scenarios).

The dialogue model resulting from the WOZ iterations is mixed-initiative. Domain communication is system-directed. *Domain communication* is communication within or about the task domain. Because of the strong limitations on active vocabulary size (see above), it was necessary during domain communication to leave the main initiative with the system. The system maintains dialogue initiative by concluding all its turns by a non-open question to the user, i.e. a question which asks for a well-defined piece of information, such as a choice between binary options, a date or time, or a destination. A field study was made of the most natural order in which to exchange the needed information. The implemented task structure conforms to the most common structure found in human-human domestic airline ticket reservation dialogues recorded in a travel agency. Whereas domain communication is system-directed, users can take the initiative in meta-communication with the system. *Meta-communication* is communication about the user-system communication itself and is usually being undertaken for purposes of clarification or repair. Whenever needed, users may initiate meta-communication to resolve misunderstanding or lack in understanding by using one of the keywords 'change' and 'repeat'. The system initiates meta-communication by saying "Sorry, I did not understand" or by asking the user, after a long pause, "Are you still there?"

In addition to contributions to meta-communication and to the achievement of particular reservation tasks, the system provides two pieces of general information: (i) the system's introduction provides information on what the system can and cannot do and how to interact with it (Figure 1). (ii) An explanation is provided of the different types of discount that are possible on return tickets. In order not to waste the time of experienced users, the system provides this information only to novice users. Figure 3 shows a dialogue from the user test of the implemented system. The dialogue is based on the scenario shown in Figure 2. The user has already made one reservation and continues without making a new call, thereby avoiding the introductory phrases shown in Figure 1. Dialogue and scenario examples have been translated from the Danish.

S1: Hello, this is the DanLuft reservation service for domestic flights. Do you know how to use this system?
U1: No.
S2a: The system can reserve tickets for Danish domestic flights. You use it by answering the system's questions. In addition you may use the two special commands "repeat" and "change" to have the most recent information repeated or changed. The system will only understand you when you answer its questions briefly and one at a time.
S2b: Please state your customer number.

Fig. 1. The introduction (S2a) to the Danish dialogue system. S means system, U means user.

Anders Bækgaard (ID-number 6), Paul Dalsgaard (ID-number 3) and Børge Lindberg (ID-number 4) work in a department in Aalborg that has customer number 3. They are all going to Copenhagen on the first weekend in February. They want to depart by the earliest flight on Saturday at 7:20 and return by the latest flight on Sunday at 22:40.

Fig. 2. The scenario T32.

3 The User Test

The user test was carried out with a simulated speech recogniser [2]. A wizard keyed in the users' answers into the simulated recogniser. The simulation ensured that typos were automatically corrected and that input to the parser corresponded to an input string which could have been recognised by our real speech recogniser. In this set-up, the recognition accuracy would be 100% as long as users expressed themselves in accordance with the vocabulary and grammars known to the system. Otherwise, the simulated recogniser would turn the user input into a string which only contained words and grammatical constructions from the recogniser's vocabulary and rules of grammar.

The test was based on 20 different scenarios which had been constructed to enable exploration of all aspects of the task structure. As the flight ticket reservation task is a well-structured task in which a prescribed amount of information must be exchanged between user and system, it was possible to extract from the task structure a set of sub-task components, such as number of travellers, age of traveller and discount versus normal fare, any combination of which should be handled by the system. The scenarios were generated from systematically combining these components.

Twelve novice subjects, mostly professional secretaries, participated in the user test. The subjects conducted the scenario-based dialogues over the telephone in their normal work environments in order to make the task as realistic as possible. The subjects were given a total of 50 tasks based on 48 scenarios. A *task* consists in ordering one or more tickets for one route. The number of recorded dialogues was 57 because subjects sometimes reiterated a failed dialogue and eventually succeeded in the task. A *dialogue* is one path through the dialogue structure. Subjects failed on seven tasks. The task transaction success rate thus was 43/50% = 86%.

Each dialogue was recorded and all transactions between the individual system modules were logged. The recorded dialogues were transcribed and analysed. The analysis aimed at detecting dialogue interaction problems and was done as follows. Based on the dialogue structure, a template was built which contained the system's questions. For each scenario, normative system questions and user answers were filled into the template. The key contents of the actual dialogues were then plotted into the template. Comparison between normative and actual system and user answers led to the identification of three major classes of interaction problems: (1) linguistic problems, (2) problems of dialogue interaction, and (3) other problems, such as system breakdown. (2) splits into

(A) dialogue design problems and (B) user errors. The dialogue design problems (A) identified in the user test are presented and discussed in [6]. The following section focuses on describing and illustrating (B).

4 User Errors

The concept of a 'user error' is a complex one, as we shall see. For now, let us define a user error as a case in which a user does not behave in accordance with the full normative model of the dialogue. In controlled user testing, the full normative model of the dialogue is defined by (i) explicit designer instructions to users, i.e. the scenarios that users have to carry out in dialogue with the system; (ii) explicit system instructions to users, such as the system's introduction to itself (Figure 1); (iii) explicit system utterances during dialogue; and (iv) implicit system 'instructions' to users. (i) through (iv) will be illustrated and discussed below. (i) is absent in field testing of systems and in practical system use. In such cases, the full normative dialogue model reduces to (ii)-(iv). (ii) is very important to SLDS design and use because SLDSs are vastly inferior dialogue partners when compared to humans. This means that humans must be told how to interact with an SLDS lest they treat the system as a human dialogue partner. Users, of course, should react appropriately to the system's utterances during dialogue (iii), given their scenarios and the dialogue context. The need to include (iv) follows from the important fact that it is impractical to explicitly inform users about all the types of behaviour they should avoid during dialogue with the system. Rather, users should be made to grasp the general fact that the system is a severely restricted dialogue partner which should be treated accordingly.

S1: Do you want anything else?
U1: Yes.
S2: Customer number 4. Is that correct?
U2: No, customer number 3.
S3: Customer number 10 does not exist. Please state your customer number.
U3: Customer number 3.
S4: Customer number 3.
U   (overlapping speech): Yes.
S4   (continued): How many people will travel?
U4: Three people.
S5: Three people. Please state the id-number of the first person.
U5: Number 6.
S6: Please state the id-number of the second person.
U6: Number 3.
S7: Please state the id-number of the third person.
U7: Number 4.
S8: Id-numbers 6, 3 and 4, Anders Bækgaard, Paul Dalsgaard and Børge Lindberg. Where does the journey start?
U8: The journey starts in Aalborg.
S9: From Aalborg. Where does the journey end?
[Continued on the next page.]

U9: In Copenhagen.
S10: In Copenhagen. Do you want a return ticket?
U10: Yes.
S11: Are you particularly interested in discount?
U11: Yes.
S12: On which date will the journey start?
U12: The first weekend in February.
S13: Friday February 10th. At which time?
U13: It must be on Saturday at 7:20. E2, E4, E6a
S14: There is no departure at 7:20. The closest other red departure is at 10:50. Do you want this departure?
U14: No.
S15: At which time?
U15: Saturday at 7:20. E6a
S16: There is no departure at 7:20. The closest other red departure is at 10:50. Do you want this departure?
U16: Yes.
S17: On which date will the return journey start?
U17: On Sunday February 5th.
S18: Sunday February 12th. At which time?
U18: 22:40. E2
S19: 22:40. You have now booked 3 tickets for Anders Bækgaard, Paul Dalsgaard and Børge Lindberg from Aalborg to Copenhagen on Friday February 10th at 10:50 returning on Sunday February 12th at 22:40. The tickets are on green conditions. Your reference number is 51. The total price is 2820 kroner. Are the tickets to be mailed or will the travellers pick them up at the airport?
U19: They should be mailed. E2
S20: The tickets will be mailed about one week before the departure. Do you want anything else?
U20: Yes.

Fig. 3. The dialogue T32a from the user test of the Danish dialogue system (translated from the Danish) [5]. The dialogue is based on the scenario shown in Figure 2. S means system, U means user. Utterances marked with an E and a number contain user errors and are discussed in more detail in Section 4.

According to the above definition of 'user error', 102 individual user errors were found in the user test corpus which contains a total of 998 user utterances. Note that a single utterance may contain several errors (cf. Figures 4, 7 and 8). Each user error was characterised with respect to its context, its nature was explained and, whenever possible, a preventive measure was proposed.

The six user errors identified in the dialogue shown in Figure 3 are analysed in Figures 4 to 9. E(n) refers to the user error typology in Table 1. ER means error, EX means explanation, PM means preventive measure. S means system, U means user. The number following each S and U refers to the dialogue in Figure 3. The dialogue was a transaction failure.

ER: S12: On which date will the journey start? U12: The first weekend of February. S13: Friday February 10th. At which time? U13: It must be Saturday at 7:20. EX: The user ignores the date fed back by the system and only tries to change Friday into Saturday. PM: People sometimes do not listen sufficiently carefully. They may also care less in experimental settings than in real life.

Fig. 4. A user error identified in the dialogue shown in Figure 3. The error is of type E2: Ignoring clear system feedback. This error was considered a direct cause of the transaction failure.

ER: S17: On which date will the return journey start? U17: On Sunday February 5th. S18: Sunday February 12th. At which time? U18: 22:40. EX: The user ignores the system feedback on date. PM: People sometimes do not listen sufficiently carefully. They may also care less in experimental settings than in real life.

Fig. 5. A user error identified in the dialogue shown in Figure 3. The error is of type E2: Ignoring clear system feedback. This error was considered a direct cause of the transaction failure.

ER: S19: You have now booked ... on Friday February 10th at 10:50 returning on Sunday February 12th at 22:40 ... at the airport? U19: They should be mailed. EX: The user ignores the system feedback on date. PM: People sometimes do not listen sufficiently carefully. They may also care less in experimental settings than in real life.

Fig. 6. A user error identified in the dialogue shown in Figure 3. The error is of type E2: Ignoring clear system feedback. This error was considered a direct cause of the transaction failure.

ER: S13: Friday February 10th. At which time? U13: It must be Saturday at 7:20. EX: The user is too occupied with the present problem to remember to use 'change' when trying to change Friday into Saturday. PM: 'Change' is not natural. Prefer more natural meta-communication.

Fig. 7. A user error identified in the dialogue shown in Figure 3. The error is of type E4: Change through comments.

ER: S13: Friday February 10th. At which time? U13: It must be Saturday at 7:20. EX: Natural user response package. PM: Allow naturally related information, such as date and time, to be provided in the same user answer.

Fig. 8. A user error identified in the dialogue shown in Figure 3. The error is of type E6: Answering several questions at a time.

ER: S15: At which time? U15: Saturday at 7:20.
EX: Natural user response package.
PM: Allow naturally related information, such as date and time, to be given in the
same user answer.

Fig. 9. A user error identified in the dialogue shown in Figure 3. The error is of type E6:
Answering several questions at a time.

A more thorough analysis of the user errors identified according to the definition above
revealed, however, that a significant number were caused by problems in the design of
the system's dialogue contributions. For instance, users responded differently from
what they should have responded according to the scenario because of missing system
feedback or because a system question was too open and invited users to respond in
ways which we had not intended. We shall ignore such cases and focus on the dialogue
errors that were made solely by the users. This leaves 61 individual user errors for
discussion in what follows.

Error Types	Error Sub-Types	No. of Cases	Preventive Measure
E1. Misunderstanding of scenario	a. Careless reading or processing	14	Use clear scenarios, carefully studied, to reduce errors.
E2. Ignoring clear system feedback	a. Straight ignorance	7	Encourage user seriousness to reduce errors.
E3. Responding to a question different from the clear system question	a. Straight wrong response	4	Encourage user seriousness to reduce errors.
	b. Indirect response	3	Disguised dialogue design problem.
E4. Change through comments (including 'false' keywords)	a. Cognitive overload	17	Disguised dialogue design problem.
E5. Asking questions	a. Asking for decision-relevant information	3	Disguised dialogue design problem.
E6. Answering several questions at a time	a. Natural response 'package'	10	Disguised dialogue design problem.
	b. Slip	1	None.
E7. Thinking aloud	a. Natural thinking aloud	1	None.
E8. Non-cooperativity	a. Unnecessary complexity	1	None.

Table 1. The initially identified user error types and sub-types.

The remaining 61 user errors are of eight different types as shown in Table 1. Two error types (E3 and E6) were divided into sub-types. E1 includes the scenario violations, i.e. violations of explicit designer instructions. E2 and E3a include cases in which users did not pay attention to explicit system utterances (feedback and questions). E3b is closely related to E5 (see below). E3b, E4, E5, E6 and E7 represent violations of explicit system instructions provided in the system's introduction (Figure 1). In E8 the user violates implicit system instructions. We will now discuss each error type in more detail.

E1. Misunderstanding the Scenario

As remarked earlier, scenario misunderstandings are artefacts of controlled user testing. Nevertheless, controlled user testing is important in systems design and it may be worth considering ways of preventing user errors in controlled test environments. It should be noted that scenario misunderstandings cannot give rise to transaction failure. The system cannot be blamed for not knowing that the user was supposed to have asked for something different from what s/he actually did ask for. Transaction failure occurs only when users do not obtain the reservation they actually ask for. In fact, scenario misunderstandings rarely lead to other forms of dialogue interaction problems. They may do so if the user mixes up several possible scenarios and thereby succeeds in providing inconsistent input. Normally, however, users just carry out a different scenario. On the other hand, this may affect system evaluation. A scenario which is not carried out may result in that part of the dialogue model remains untested.

Almost one fourth of the 61 user errors were due to users acting against the instructions in the scenarios. These errors were of three (task-dependent) kinds: (a) users asked for one-way tickets instead of return tickets; (b) users were not interested in discount although according to the scenario they should have been; and (c) users tended to miscalculate the date of departure if only given indirectly in the scenario. It seems likely that the main reason for the many scenario misunderstandings is the artificial experimental situation. People care less in an experiment than they do in real life and therefore tend not to prepare themselves sufficiently for the dialogue with the system. In addition, unclear scenarios cause errors. E1 thus raises two issues in the preparation of controlled user testing: (i) to reduce the number of errors, scenarios should be made as clear as possible. Nothing is gained by unclear or misleading scenarios. Clear scenarios should not be confused with *simple* scenarios. Scenarios should reflect the types of information real users actually have when addressing the system. This information may be complex and some scenarios should reflect that. This means that users may have to perform some mental processing of the scenario information in order to provide correct answers to the system's questions. (ii) Users should be encouraged to carefully prepare themselves on the scenarios they are to complete in conversation with the system. This should mirror the interest real users have in getting the system to deliver what they want. A practical solution is to promise an award to subjects who stick to their scenarios in conversation with the system. Awards depend on culture so we will not suggest a good bottle of wine as the sole solution.

Whatever preventive measures are taken, however, scenario misunderstandings are not likely to be totally absent from controlled user tests but reducing their number is an important goal.

E2. Ignoring Clear System Feedback

The speech recognition capabilities of most telephone-based systems are still fragile. It is therefore important that users listen carefully to the system's feedback to verify that they have been correctly understood. Of the seven transaction failures in the user test, one was caused by a combination of a dialogue design problem and a user who ignored clear system feedback. A second transaction failure occurred solely because the user did not pay sufficient attention to the system's feedback which made it clear that the user had been misunderstood (Figures 3, 4, 5 and 6). Three of the seven detected E2 cases occurred in this dialogue in which the user continuously ignored the system feedback on dates (Figures 4, 5 and 6). Thus, four out of the seven detected cases of ignored system feedback had severe implications for the success of the transaction. Moreover, had the user test included a real recogniser, more cases of system misunderstanding would have occurred and hence more cases in which users would have had to identify system misrecognitions from the system's feedback.

E2 raises the issue of encouraging test subjects to 'act' seriously in dialogue with the system and be very attentive to what the system says because recognition in SLDSs is much more error-prone than the hearing capabilities of normal humans. This would help reducing the number of user errors caused by their ignoring system feedback. Nothing is gained by having subjects who care too little about what is going on during the dialogue. Whatever preventive measures are taken, however, the problem of user inattentiveness is not likely to completely go away. This is true of both 'artificial' user tests and real-life use of commercial systems.

The notion of a transaction failure that is caused by a 'clean' user error may be controversial. It might be argued that transaction failures should be caused by systems design errors of one kind or another. On the other hand, it might be said that most user errors of ignoring clear system feedback only arise because the system has misunderstood the user in the first place. This problem does not seem to have any obvious solution. Whatever one chooses to do, this should be made clear in the definition of 'transaction failure' adopted because the resulting transaction failure percentage constitutes an important quantitative measure of system performance.

E3. Responding to a Question Different from a Clear System Question

E3 has at least two sub-types. The first sub-type, E3a, included four cases in which users gave a straight wrong response to a system question, for instance by answering "Saturday" to the question about departure airport. In one case the answer was not understood by the system and in three cases it was misunderstood. E3a raises the same issue as did E2 of encouraging users to seriously pay attention to the system's utterances. Similarly, E3a errors are not likely to go away completely, neither in 'artificial' user tests nor in real-life interaction.

The second sub-type, E3b, concerns *indirect* user responses. For instance, a user answered "it must be cheap" to the question of hour of departure. In human-human conversation, indirect answers of this type would be perfectly all right. An indirect response indicates that the speaker does not possess the information necessary to provide a direct answer. In response to the indirect user answer quoted above, a human travel agent would list the relevant departures on which discount may be obtained. Our SLDS, however, has limited inferential capabilities and is not able to cope with indirect responses. They will be either not understood or misunderstood.

E3b is among the most challenging types of user errors in the test material. Indirect responses are natural to humans in situations in which they do not have sufficient information to produce a direct response. In such cases, we provide instead the information that we actually possess, leaving it to the interlocutor to infer the information asked for. We do this cooperatively, of course, only in cases in which the interlocutor can be assumed to have the information needed to perform the inference. The system, posing as a perfect domain expert, may legitimately be assumed to possess the required information. What the user overlooks, however, is that the system does not have the *capability to draw the proper inferences* from the user's information. The E3b cases, therefore, raise the hard issue of the extent to which dialogue designers should consider providing their system with the appropriate inferential skills. There does not currently appear to exist a principled answer to this problem. Furthermore, it may be argued that indirect user responses are not user errors at all. They do not conflict with the system's introduction (Figure 1). At best it might be argued that indirect responses conflict with the difficult requirement on users which we have called 'implicit instructions' to users (see above). If, however, we are right in the above interpretation of E3b-type user contributions, they are in fact oblique questions asking for information (see E5 below). We shall return to E3b in the concluding discussion.

E4. Change through Comments

E4 gave rise to numerous (almost 30%) user errors in the test. In 16 out of 17 cases, users tried to make corrections through natural sentences rather than by using the keywords prescribed in the system's introduction (Figure 1). An example is shown in Figure 4. In none of these cases was the requested correction understood as intended. Only in one case did the user achieve the intended correction. In this case, the user used a keyword different from 'change' but meaning the same, which accidentally was recognised as 'change'. The theoretical importance of these findings is that of emphasising the undesirability of including designer-designed user keywords in dialogue design for SLDSs. Such keywords will neither correspond to the keywords preferred by all or most users nor to the natural preference among native speakers to reply in spoken sentence form rather than through keywords. It is furthermore our hypothesis that the more cognitive load a user has at a certain stage during dialogue task performance, the more likely it is that the user will ignore the system's instructions concerning the specific keywords to be used.

E4 raises the hard issue of allowing users a more natural form of repair meta-communication.

E5. Asking Questions

E5 is among the most challenging types of user errors in the test material and is closely related to E3b (see above). Like the E3b cases, the E5 cases all occur when the system has asked for an hour of departure. For instance, a user then asked "what are the possibilities?". What the observed cases show is that reservation dialogue, in its very nature, so to speak, is *informed reservation* dialogue. It is natural for users who are making a reservation or, more generally, ordering something, that they do not always possess the full information needed to decide what to do. In such cases, they ask for the information. Since the system poses as a perfect domain expert, this is legitimate. What users overlook, however, and despite what was said in the system's introduction (Figure 1), is that the system does not have the skills to process their questions. As with E3b above, it is not clear what the dialogue designer should do about this problem in the short term. Current systems are not likely to be able to understand all possible and relevant user questions in the context of reservation tasks. The optimistic conclusion is that E3b and E5 only constituted 4 user errors in total in the user test, and that skilled users of the system will learn other ways of eliciting the system's knowledge about departure times. However, a principled solution to the problem only seems possible through enabling the system to conduct rather sophisticated mixed-initiative domain dialogue.

E6. Answering Several Questions at a Time

E6 has at least two sub-types. The first sub-type, E6a, gave rise to many (about 16%) user errors in the test. Examples are a user who answers "the journey starts on Friday at 8:15" when asked for a date of departure, and a user who answers "no, change" when asked if it is correct that the destination is Karup. Other examples are shown in Figures 8 and 9. In 7 of the 10 cases, only the part of the user's response which answered the system's question was understood. In the remaining 3 cases the entire user response was misunderstood. What this error type suggests is that (i) users naturally store information in 'packages' consisting of several pieces of information. This means that they are unlikely to consistently split these packages into single pieces of information despite having been told to do so in the system's introduction (Figure 1). Dialogue designers should be aware of the existence of such natural information packages and enable their system to understand them. (ii) Users have stereotypical linguistic response patterns, such as prefixing a 'change' keyword with a 'no'. Dialogue designers should be aware of these natural stereotypes and enable the system to understand them. This problem appears solvable by today's technology. Our SLDS is already able to accept such stereotypes in several cases, such as when information on departure and arrival airports is being provided in the same utterance. However, due to the present, strong limitations on active vocabulary we have not been able to allow natural information packages and stereotypes throughout the reservation dialogue.

The second sub-type, E6b, illustrates a phenomenon which no feat of dialogue design is likely to remove, i.e. the naturally occurring slips-of-the-tongue in spontaneous speech. Slips do not appear to constitute any major problem, however. Only one slip causing an interaction problem occurred in the entire corpus: when asked for the customer number, the user said "four, no sorry, change, change". Only the number was recognised forcing the user to change it in the following utterance.

E7. Thinking Aloud

E7 illustrates another phenomenon which no dialogue design effort is likely to remove, i.e. the naturally occurring thinking-aloud in spontaneous speech. Thinking-aloud does not appear to constitute a major problem, however. Only one case of natural thinking-aloud occurred in the entire corpus: when asked for the hour of departure, the user said "well, let me see, at 8:30 at the latest".

E8. Non-Cooperativity

E8 illustrates yet another phenomenon which cannot be removed through dialogue design, i.e. the deliberately non-cooperative user. Only one case of deliberate user non-cooperativity was detected in the test corpus. The user replied "the ticket should not be sent" to the system's question of whether the ticket should be sent or would be picked up at the airport. This reply would not have been considered non-cooperative if produced in human-human conversation. However, the reply is unnecessarily complex and cannot be handled by our SLDS. We know that the particular user who caused the problem was deliberately testing the hypothesis that the system would be unable to handle the input because she said so in the telephone interview following her interaction with the system. SLDSs designers have no way of designing dialogues with sufficient robustness to withstand deliberately non-cooperative users. Nor should SLDSs designers attempt to do so, apart, of course, from ensuring that the system will not break down and that deliberately non-cooperative users cannot cause any harm. The simple fact is that deliberately non-cooperative users, when successful, will fail to get their task done.

5 Conclusion

The E1 errors are of only minor importance as they will disappear when the system is being used in real life. Furthermore, the evidence suggests that E1 errors do not tend to cause severe dialogue interaction problems. Similarly, E8 errors are of minor importance because users will stop experimenting with the system when they want the task done. E6b and E7 can hardly be prevented but, at least according to our test material, they are infrequent and do not cause severe problems of interaction.

E2 and E3a seem to have a much larger effect on dialogue transaction success. Although they can hardly be completely avoided, it is likely that their number can be reduced by clearly making users aware of the importance of paying attention to system feedback and system questions. Real-life users are likely to be more attentive.

E3b, E4, E5 and E6a are the most challenging user error types found in the corpus. They would all be perfectly acceptable in human-human dialogue. However, because of the limited dialogue capabilities of our SLDS, it is clearly stated in the system's introduction how users should interact with it in order to prevent these errors. Whereas E3b is less clear (see Section 4 above), the E4, E5 and E6a errors all violate the system's explicit instructions. The important question is why so many users violate exactly these instructions. A likely explanation is that, at least for many users, it is not *cognitively feasible* to follow the system's explicit instructions. In an extreme example: had we asked users to always use exactly four words in their responses to the system's questions, this would clearly have been cognitively infeasible. Similarly, several of the things which the system's introduction asks users to do or avoid doing turn out to be unrealistic given the dialogue behaviour that is natural to most people. This reveals a fundamental shortcoming in our initial concept of 'user error' (Section 4). It is not sufficient to provide clear and explicit instructions to users on how to interact with the system. *It must also be possible for users, such as they are, to follow these instructions in practice.* The conclusion is that E3b, E4, E5 and E6a are *not* user errors at all but rather constitute more or less difficult problems of dialogue design.

E3b, E4, E5 and E6a are otherwise very different. E3b and E5 result from a mismatch between generic task type (ordering) and the type of dialogue initiative adopted for the application (system-directed domain communication). E4 and E6a belong to a much more general class of human-machine interaction problems. For years, in fact, experts on human error in the field of human factors have been aware of the broad category of errors illustrated by E4 and E6a. The reason why they are easy to overlook during design and until the user and field test data come in, is that, *in principle,* we can all avoid them. For instance, we can all easily say 'change' when we want to correct a system misunderstanding. During actual task performance, however, whether the task be one of driving a car or communicating with an SLDS, we tend to fall back on our natural skills and what is inherent to the human cognitive processing architecture, more or less ignoring rules or instructions that conflict with those skills and that architecture.

What are the implications of the findings reported in this paper? We emphatically do not want to argue that, because of problems such as E3b, E4, E5 and E6a, SLDSs of the same general type as ours cannot be used in realistic applications. None of these problems caused transaction failure. The E6a problems can be removed using current dialogue design techniques (Section 4). The E3b and E5 problems were few. And the E4 problems, of which there were many, might at least be reduced in number through a larger active vocabulary. Of course, user-system interaction can and should be improved through many other means than those addressing the occurrence of user errors. We have discussed such means in our analysis of the dialogue design errors that were identified in the user test corpus [6]. The present paper has focused on how and to what extent user errors can be prevented.

Acknowledgements

The Danish dialogue system was developed in collaboration between the Center for PersonKommunikation at Aalborg University (speech recognition, grammar), the Centre for Language Technology, Copenhagen (grammar, parsing), and the Centre for Cognitive Science, Roskilde University (dialogue and application design and implementation, human-machine aspects). The project was supported by the Danish Research Councils for the Technical and the Natural Sciences. We gratefully acknowledge the support.

References

[1] H. Aust, and M. Oerder, Dialogue Control in Automatic Inquiry Systems, *Proceedings of the ESCA Workshop on Spoken Dialogue Systems*, Vigsø, 121-124, 1995.

[2] N.O. Bernsen, H. Dybkjær and L. Dybkjær, Exploring the Limits of System-directed Dialogue. Dialogue Evaluation of the Danish Dialogue System, *Proceedings of Eurospeech '95*, Madrid, 1457-60, 1995.

[3] R. Cole, D.G. Novick, M. Fanty, P. Vermeulen, S. Sutton, D. Burnett and J. Schalkwyk, A Prototype Voice-Response Questionnaire for the US Census, *Proceedings of the ICSLP '94*, Yokohama, 683-686, 1994.

[4] H. Dybkjær, N.O. Bernsen and L. Dybkjær, Wizard-of-Oz and the Trade-off between Naturalness and Recogniser Constraints. *Proceedings of Eurospeech '93*, Berlin,. 947-50, 1993.

[5] L. Dybkjær, N.O. Bernsen and H. Dybkjær, Evaluation of Spoken Dialogues. User Test with a Simulated Speech Recogniser. *Report 9b from the Danish Project in Spoken Language Dialogue Systems*. Roskilde University, February 1996. 3 volumes of 18 pages, 265 pages, and 109 pages, respectively.

[6] L. Dybkjær, N.O. Bernsen and H. Dybkjær, Reducing Miscommunication in Spoken Human-Machine Dialogue. *Proceedings of AAAI "96 Workshop on detecting repairing and preventing human-machine miscommunication*, Portland, 1996.

[7] W. Eckert, E. Nöth, H. Niemann and E. Schukat-Talamazzini, Real Users Behave Weird - Experiences Made Collecting Large Human-Machine-Dialog Corpora, *Proceedings of the ESCA Workshop on Spoken Dialogue Systems*, Vigsø, 193-196, 1995.

[8] N.M. Fraser and G.N. Gilbert, Simulating Speech Systems, *Computer Speech and Language 5*, 81-99, 1991.

Towards a Dialogue Taxonomy

Nils Dahlbäck

Department of Computer and Information Science, Linköping University, S-581 83
Linköping, Sweden, nilda@ida.liu.se

Abstract. Two interrelated points are made in this paper. First, that
some of the characteristics of the language used in spoken dialogues are
also observed in typed dialogues, since they are a reflection of the fact
that it is a dialogue, rather than the fact that it is spoken. A corollary
of this is that some of the results obtained in previous work on typed dialogue, especially Wizard of Oz simulations of human-computer natural
language dialogues, can be of use to workers on spoken language dialogue
systems. Second, it is claimed that we need a more fine grained taxonomy of different kinds of dialogues. Both for making it possible to make
use of results obtained by other workers in developing dialogue systems,
and for the development of our theoretical understanding of the influence
of non-linguistic factors on the language used in human-computer dialogues. A number of such dimensions are described and their influence
on the language used is illustrated by results from a empirical studies of
language use.

1 Introduction

There is a strong trend in present-day computational linguistics towards empirically based models and theories. In the area of dialogue systems there is
concurrently a shift away from typed dialogue to spoken dialogue systems. Since
there exist a body of empirical results and models for typed dialogues systems,
the question arises to which extent these results are useful and applicable also
for spoken dialogue systems. What makes this question difficult to answer is the
relative neglect of mainstream linguistics concerning the issue of what characterizes the language used in different domains and situations[1]. Consequently, many
of the views on what characterizes spoken or written language are often less than
accurate for professional linguists and laypersons alike. Linell (1982) has force-
fully argued that mainstream linguistics suffer from a "Written Language Bias".
And Volosinov (1973:71) claims that "European linguistic thought formed and
matured over concern with the cadavers of written languages; almost all its basic
categories, its basic approaches and techniques were worked out in the process
of reviving these cadavers".

While not all workers on spoken language and dialogue would subscribe to
the strong positions taken by Linell and Volosinov, there seem to be an explicit or

---

[1] In some areas of linguistics this is of course not a correct characterization, as illustrated by the work on sub-languages by Grishman and Kittredge (1986) and others.

implicit view among many workers in the field that large portions of the current linguistic theories are not relevant for the development of spoken dialogue systems. While not explicitly stated, the implicit argument seems to be something like the following.

Premise 1. Spoken and written language are different.

Premise 2. Current linguistic theories (e.g. grammars) are developed for written language.

Conclusion: We cannot make use of results and theories based on written language, but have to start more or less from scratch.

While I agree with the basic arguments by Linell and Volosinov, and while I agree that there are important differences between spoken and written language, one claim of the present paper is that the dialogue community risks throwing out the baby with the bathwater, and this for the following reason. The arguments on the differences between spoken and written language do not sufficiently take into account the fact that the prototypical spoken and written language situations differ on more than one dimension; the prototypical spoken language is a dialogue, the prototypical written language is a monologue. And the claim here is that some of the characteristics attributed to spoken language does not depend on the fact that it is spoken, but that it is a dialogue.

Before continuing I want to stress that I do not deny that there *are* differences between spoken and written dialogues. The speech recognition problem of decoding the acoustic signal is an obvious example of this. But since these differences are well known, my aim is here instead to argue that there are important similarities between dialogues, regardless of the communication medium.

2 Spoken and Written Dialogues

There is no denying that the language used in naturally occurring dialogues does not look like the language we were taught to write in grammar school. And there is no denying that the language used in the prototypical spoken dialogue and in the prototypical written text (monologue) differ along the lines often described. But prototypes are not the only members of categories. The issue I want to address here is to what extent it is the medium (spoken versus typed) or the interaction type (dialogue versus monologue) that accounts for the linguistic qualities observed in spoken dialogue. Unfortunately there are not many published studies that I am aware of that explicitly address this issue. But there have been quite a number of studies of typed dialogues, many being so-called Wizard of Oz-experiments. While the latter fact perhaps makes it difficult to ascertain whether the observed qualities are caused by the interaction modality or the perceived qualities of the non-human dialogue partner, this does not need to be a major concern for the computational linguistic community, since the aim here presumably is to design interfaces for interaction with computers.

There are a number of studies of human-computer interaction using typed natural language, that show that these dialogues exhibit many of the attributes characteristic of spoken dialogues. If we look at some of the early ones, Reilly

(1987a 1987b) and (Malhotra 1975) have shown that the syntactic variation is rather limited. Furthermore, so-called ill-formed input is very frequent, especially the use of fragmentary sentences and ellipsis. Guindon et al (1986, 1987) present a detailed analysis of the language used by their subjects. They report that 31 % of the utterances contained one or more ungrammaticalities. The most common of these were fragments (13 %), missing constituents (14 %), and lack of agreement (5 %).

These results (and others, for a larger review of the early studies in the field, see Dahlbäck, 1991) indicate that typed and spoken dialogues share many qualities, and in many respects deviate from ordinary typed language in similar ways. Hence, the argument from the Call for Papers for a recent workshop on Spoken Dialogue Systems that spoken dialogues differ from typed ones because

" ... spoken input is often incomplete, incorrect and contains interruptions and repairs; full sentences occur only very occasionally. Therefore new basic units for the development of dialogue models have to be proposed ... "

seems to be somewhat lacking in empirical support.

There are of course also observed differences between typed and spoken dialogues. Cohen (1984) studied the effects of the communication channel on the language used in task oriented dialogues. When comparing spoken (telephone) and teletype conversations he noted that " keyboard interaction, with its emphasis on optimal packaging of information into the smallest linguistic "space", appears to be a mode that alters the normal organization of discourse". (Cohen, 1984, p 123) To take one example, the use of cue-words to introduce new discourse segments occurs in more than 90 % of the cases of spoken discourse, but in less than 45 % of the written dialogues.

But note here that the observations made by Cohen and others, do not indicate that the language of the typed dialogues resembles the prototypical written language, but instead that it in some respects, e.g. tersness, deviates even more from the norm of 'normal' written language than does the language of spoken dialogues.

I am not familiar with any systematic studies of the differences in dialogue structure between spoken and typed dialogues. What seems to affect the dialogue structure is rather whether you are interacting with a computer or a person. I will get back to this issue later.

While the number of studies is limited, and while therefore the conclusions drawn need to be considered tentative, at least to my mind it seems clear that there are a number of important similarities between spoken and typed dialogue, and that consequently workers on spoken dialogue, at least to some extent, can make use of the results obtained from analyzing typed dialogues.

While I have previously stressed the similarities between spoken and typed dialogues, it should also be noted that there seem to be some differences too, for instance regarding the dialogue structure. An illustration of this is is found in the different kinds of basic dialogue structure proposed by us for typed dialogues

(Dahlbäck 1991, Dahlbäck & Jönsson 1992, Jönsson 1993) and for Bilange (1991) for spoken dialogues. The dialogues involve in both cases information retrieval. The dialogue models developed by us and by Bilange are very similar, whichs makes a comparison between the models on a detailed level possible. We then find that the spoken dialogues seem to exhibit a three-move structure (called Negotiation, Reaction, Elaboration by Bilange), whereas in the typed dialogues a two-move structure (Initiative, Response) is sufficient. Since also Stubbs (1983) found a similar three-move pattern in his analysis of spoken dialogues, this suggests that there is some general difference between spoken and typed dialogues in this respect.

Before leaving this dimension I wish to suggest that one important difference between spoken and typed dialogues with computers affecting the discourse, is that in typed dialogues parts of the dialogue remain in front of the user when planning and executing the next move. We have for instance found in our work on typed dialogues that even with extremely long response times (due to a very slow simulation environment at the time), users make use of anaphoric expressions, including ellipsis in the dialogues.

3 Towards a Dialogue Taxonomy

The results from the studies described above indicate that there are important similarities between dialogues, regardless of the medium used, but these studies also make it clear that the four categories of dialogue created by the two binary dimensions, spoken-written and monologue-dialogue, are not sufficient for describing the factors that influence the language used in aspects important for the development of computational models of discourse. The rest of this paper will therefore be devoted to providing the first steps towards the development of such a taxonomy, with special emphasis on discourse aspects relevant for computational theories of discourse[2].

As pointed out above, the difference between the prototypical spoken and written language is really not one but many. Rubin (1980) suggests that the communicative medium or the communication channel, should be partitioned into the following seven dimensions: modality (written or spoken), interaction, involvement, spatial commonality, temporal commonality, concreteness of referents (whether objects and events referred to are visually present or not), separability of characters. Below I will present observations suggesting that at least some of these dimensions influence language in aspects of interest to computational linguists.

I make no claim that the dimensions described below constitute an exhaustive list, nor do I wish to claim that they are independent. It is much too early to make such conclusions. My goal is a more modest one; I hope to initiate a discussion on some of the aspects I consider important here, since I believe that the healthy development of the field requires clarification on these issues.

---

[2] Parts of this material was previously published in Dahlbäck (1995)

In Linköping we have recently been involved in a project aimed at comparing different kinds of computational discourse models empirically. We have not only used our own dialogue corpora from previous work, but have tried to gain access to other corpora as well. We found in the course of this work that different kinds of computational models seemed to be more adapted to some kinds of dialogues than to others. This led us to partly reformulate the aims of the project to also focus on a descriptive scheme of different kinds of dialogues. No claim of originality is made concerning the dimensions mentioned here. As will be obvious to many readers, much of what is presented below is based on or influenced by work of others, probably even more so than is made evident in the references. I have tried to enforce my argument that these factors need to be taken seriously by the computational discourse community by illustrating the possible ways in which the factors mentioned influences or might influence the computational treatment of discourse.

3.1 Kinds of Agents

The fact that the dialogue partner is a person or a computer seems to influence a number of linguistic aspects, from the use of pronouns to the dialogue structure. Guindon (1988) showed that the dialogue structure differed between dialogues with persons and with computers in similar situations. When interacting with a computer, the dialogue structure was simpler. The work by Kennedy, Wilkes, Elder and Murray (1988) showed that the language used when communicating with a computer, as compared with a person in a similar situation, has the following characteristics: Utterances are shorter, the lexical variation is smaller and the use of pronouns is minimized. The results concerning the limited use of pronouns when communicating with computers has been established in a large number of studies (for a summary of a number of studies on this and other aspects of 'computerese', see Dahlbäck 1991, ch 9). In most cases it is, however, not possible to ascertain whether the differences found are caused by influences of the channel (typed vs spoken) or the perceived characteristics of the dialogue partner (human vs computer).

In a current project in Linköping we are comparing the language used when communicating with a computer or with a person in identical situations (typed information retrieval with or without the possibility of also ordering the commodities discussed). The only difference between the two situations is what the subjects are told they are interacting with; whether it is a person or a computer. In all other respects the situations are similar (and the 'wizards' are not told beforehand under which condition the specific subject is run). It is interesting to note that it is in this situation rather difficult to find any differences between the dialogues with humans and those with computers. If this result holds after a more thorough analysis, this indicates that *communication channel* and *kinds of tasks* influence the dialogue more than the perceived characteristics of the interlocutor. It is, however, still possible that there are differences between these dialogues in for example the dialogue structure, something which has not been analyzed yet.

We have recently extended this work by comparing spoken dialogues with computers from the Waxholm-project (Bertenstam et. al., 1995) with dialogues between people. In both cases the task was information retrieval. This far, our observations suggest that it is possible to adapt our LINLIN dialogue model (Dahlbäck and Jönsson, 1992), which was developed for typed dialogues with computers, to cover also spoken dialogues with computers without any major modifications (Jönsson, 1996), but this is less so for dialogues between people.

A tentative preliminary conclusion possible to draw from the collected observations mentioned above on the dialogue structure is that typed dialogues with computers, both in information retrieval and in at least some advisory situations (Guindon), exhibit a rather simple structure possible to model with a context free dialogue grammar. This is also true for spoken dialogues with computers, at least for information retrieval (Bilange), whereas the dialogue structure in dialogues between people, also in information retrieval dialogues, exhibit a more complex structure.

3.2 Interaction

The interaction dimension (dialogue versus monologue) seems to influence among other things the use of pronouns and the pattern of pronoun-antecedent relations. In typed human-computer dialogues pronouns are rarely used (Guindon, 1988, Dahlbäck & Jönsson, 1989, Kennedy et al, 1988). The anaphor-antecedent relations seem to be of a rather simple kind in these kinds of dialogue. To take one example, we found in an analysis of these patterns in one of our corpora of Wizard of Oz-dialogues that in those cases where the personal pronouns had an antecedent, the distance between pronoun and antecedent was very small. The analysis suggested that the antecedent could be found using a very simple algorithm which basically worked backwards from the pronoun and selected the first candidate that matched the pronoun on number and gender and which did not violate semantic selection restrictions (Dahlbäck, 1992). The algorithm described and evaluated on a number of computer manuals by Lappin and Leass (1994) is more complicated and uses among other things an intrasentential syntactic filter for ruling out anaphoric dependence of a pronoun on an NP on syntactic grounds. It is not clear that such a filter would improve the recognition of the antecedent in our dialogues, where instead the dialogue structure was needed to stop the search for antecedents to the pronouns when these were not found within the local structure unit. The reason for this rule was that in our corpus as many as 1/3 of the personal third person pronouns lacked an explicit antecedent, but instead made use of some kind of associative relation to the antecedent, or belonged to the class of pronouns called 'propositional' by Fraurud (1988).

As already pointed out, an important difference between spoken and typed language is that spoken language seems to contain more ungrammaticalities. But as an aside, perhaps some caution could be used here in labeling these occurences 'ungrammatical', since our grammars might be carriers of the 'Written Language Bias' that Linell (1982) talks about. Violations of number agreement is of course ungrammatical, but the use of sentece fragments seems less so, it we do not take the prototypical written language as the norm.

3.3 Shared Context

Spatial and *temporal commonality* also seem to influence aspects of discourse. Not only is the use of deictic expressions made possible with a shared temporal/spatial context, but it is also possible that the use of other anaphoric devices is influenced. Guindon (1988) found, for instance, in her analysis of advisory dialogues for the use of a statistical computer package that pronouns either had their antecedent in the current sub-dialogue, or they referred to the statistical package that was present on the screen all through the dialogue. And as an aside, it is perhaps worth pointing out that the celebrated example from Grosz' dissertation (Grosz 1977, p 30), where the pronoun 'it' is used to refer to the pump just assembled, which has not been mentioned for 30 minutes and 60 utterances, could be seen as belonging to this category too. But also in other kinds of discourse where there is no shared physical context, and where the interaction is minimized there sometimes occur privileged entities that can be referred to using a pronoun even if the antecedent in the strict sense has not been mentioned for a long time. These so-called 'primary referents' (Fraurud, 1988) are for instance the main actors in a novel.

The other dimensions discussed by Rubin are probably also important not only for human dialogues, but also for human-computer dialogues. They seem to be of use, for example, when discussing and comparing different kinds of multi-media or multi-modal interaction.

Rubin also discusses a number of message-related dimensions (without claiming them to be independent), especially topic, structure and function. I will here address two dimensions closely related to the ones mentioned by Rubin, namely task structure and kinds of shared knowledge.

3.4 Dialogue-task Distance

That the task structure influences the dialogue structure was an important aspect of Grosz' (1977) early work. But she also pointed out that for man-computer dialogues "there seems to be a continuum (...) from the totally unstructured table filing dialogues to the highly structured task dialogues (ibid, p 33). In the task oriented dialogues the structure of the task was shown to influence the structure of the dialogue and this result was the starting point for the use of the underlying task structure in the analysis of discourse. An important corollary of this is of course that the dialogue structure will differ depending on which kind of task is being performed.

But it is my impression that not only different tasks will influence the structure of the dialogue, but that this is true also for a different but closely related factor which I call *dialogue-task distance*. This is based on the observation that there seems to be a closer connection between task and dialogue in for instance an advisory dialogue than in an information retrieval dialogue. For the task oriented dialogues, we know that we need to understand the non-linguistic tasks and acts to be able to understand and respond correctly to our dialogue partner's speech acts. But this is less true for simple iformation retrieval. For instance, to answer the question of when there are express trains to Stockholm within the next two hours, in most cases there seems to be no need to know why the questioner needs to know the answer. I am not denying that there are cases when the information provider can be more helpful when knowing this. But the prime case of this is probably when it is not possible to provide an answer, as for instance when in the case above, there are no trains of the requested kind within the specified time limit. In such cases humans often seem to ask for the information needed to provide additional help and presumably computer systems can do the same.

My hypothesis is therefore that we here have a dimension which is characterized both by differences on the need to understand the underlying non-linguistic task, and on the availability of linguistic information required for doing so. Reichman (1985, p 21) has argued that "though it is true that conversational moves frequently reflect speaker's goals, it is important to stress that these moves can be identified and interpreted without reference to a speaker's underlying intent for an utterance." My claim here would be that this is true, but only for some kinds of dialogues, namely those resembling information retrieval dialogues in having a long dialogue-task distance. For other kinds of dialogues, where this distance is shorter, I am inclined to believe that Reichman's claim is less valid. And this in fact to the extent that I hypothesize that different kinds of computational discourse models are to be preferred depending on the value taken on this dimension.

The closer the language-background task connection, the more appropriate become plan or intention based models. In these situations it is less difficult to infer the non-linguistic intentions behind a specific utterance from knowledge of the general task structure and from observations on the on-going dialogue. But with larger distance between the dialogue and the underlying task, as in the information retrieval case, the more difficult it becomes to infer the underlying intentions from the linguistic structure, and at the same time the need for this information in order to provide helpful answers diminishes. In these cases dialogue grammar models based on the conversational moves or speech acts made by the interlocutors are probably a better choice.

One observation from our on-going work that seems to support this position is that we have found that the coding of the underlying intentions in an information retrieval dialogue becomes really difficult if the coding is done move by move, i.e. when the move is classified without knowledge of what follows later in the dialogue. But this is, of course, the situation a computer system will be in. A coding scheme based on more surface-oriented criteria seems to be at advantage in this situation.

It is not only the connectedness between the linguistic and the non-linguistic task that influences the complexity of the dialogue. The *number of different tasks* managed linguistically is another such factor. In our work we have compared cases of information retrieval dialogues with dialogues in the same domain (travel information) where the user also can order a ticket. In the latter case not surprisingly, a more complex topic management was required (Jönsson, 1993, Ahrenberg, Dahlbäck, Jönsson, forthcoming).

This dimension seems to me to point to an important difference between human dialogues and human-computer dialogues, since there are fewer different things that can function as topic in a dialogue with a computer system.

3.5 Kinds of Shared Knowledge

The influence of different kinds of shared knowledge between dialogue participants on the use of referring expressions has been discussed by Clark and coworkers in a number of important papers (e.g. Clark & Marshall, 1981; Clark & Carlsson, 1981; for a summary see Clark, 1985). The basic point of this work is that a necessary pre-requisite for the successful use of a definite description is that speaker and listener share a common ground of mutual knowledge, beliefs, and assumptions, and furthermore that, were it not for a number of heuristics used by people, the acquisition of this mutual knowledge would require checking an infinite number of assumptions. The bewildered or skeptical reader of this claim is referred to the original sources. In this context I only want to use Clark's taxonomy of the basic classes of such heuristics for my present purposes. Clark's claim is that there are three basic such classes or kinds of information that can be used to infer the common ground between speaker and listener; shared perceptual, linguistic and cultural knowledge. Two of these have in different ways already been addressed previously. *Perceptual knowledge* is usable when the physical or visual context is shared; the shared *linguistic knowledge* is in this context another name for the shared knowledge of the previous text or dialogue. But what has not been discussed previously is the use of shared *cultural knowledge*, where 'cultural' here is used in its widest possible sense, including factual knowledge etc.

The basic idea here is that there are things that everybody in a community knows and which therefore can be used as common ground. The problem with this is, of course, to determine if my dialogue partner belongs to the same community as I do, or rather which cultural knowledge from different sub-communities that I can assume that we share. In face-to-face communication between people, we often can make a good guess at what cultural knowledge we share with our interlocutors, since choice of clothing, ways of speaking etc give clues on this. And even though we all know of embarrassing situations when the assumptions we made were blatantly wrong, the very fact that we remember these situations as something that stands out in our memory also indicates how good we often are at making these kinds of inferences.

This seems to be an aspect worth considering when selecting tasks and domains for which an interactive computer system should or could be designed. Note that in many cases the computer is worse off than a human in the same situation not only because the computer's inferential abilities are less powerful than those of the person, but also because it has a more impoverished empirical basis to build its deductions on. It cannot see its interlocutor and does not remember the person from previous encounters.

My suggestion here is that it will be difficult, at least in the short run, to develop dialogue systems for those kinds of applications where the common cultural ground needs to be acquired during the on-going dialogue. And a possible explanation for the successful information retrieval systems developed is that they operate in domains where it can be assumed that all users will have the same basic knowledge of the domain. Hence the need for clarification sub-dialouges is diminished or obsolete, as well as the need for user-modeling of a kind not yet achieved.

4 Summary

"That language varies according to the situation is a truism; however, the details and implications of that truism are far from obvious, whether your enterprise is theory formation or system construction" (Pattabhiraman, 1994). I have here tried to show that while there is still lots of work that need to be done to make the implications of this truism obvious, there are already enough observations available to make it possible to make the first steps on the path towards this goal. I have here described a number of such dimensions that I believe influence one or more important parameters for any kind of computational theory of discourse, and have tried to illustrate their possible influence on different discourse phenomena. The dimensions discussed were the following.

- Modality (spoken or written)
- Kind of agent (human or computer)
- Interaction (dialogue or monologue)
- Context (spatial and/or temporal commonality)
- Number and types of tasks
- Dialogue-task distance
- Kinds of shared knowledge (perceptual and/or linguistic and/or cultural)

The list is by no means intended to be all-inclusive and final. There are in all probability other dimensions not mentioned here that are of equal importance. And even for those dimensions described here, we do not know enough about their influence of language and interaction. But I believe that we do know enough to take them seriously in our continued work on empirically based computational dialogue models.

5 Acknowledgements

The ideas presented in this paper have evolved during a number of years when I have been involved in the development of a natural language interface for Swedish at the Natural Language Processing Laboratory at the Department of

Computer and Information Science, Linköping University. I greatfully acknowledge the inspiring discussions, help, and critique from the members of the group, and especially Lars Ahrenberg and Arne Jönsson.

This work was supported by the Swedish Research Council for Research in the Humanities and Social Sciences (HSFR).

References

1. Ahrenberg, L., Dahlbäck, N., and Jönsson, A. (Forthcoming). Dialogue mangagement for natural language interfaces. *Manuscript in preparation.*
2. Bertenstam, J., Blomberg, M., Carlson, R., Gustafson, J., Hunnicutt, S., Högberg, J., Lindell, R., Neovius, L., de Serpa-Leitao, A., Nord, L., and N. Ström (1985) Spoken dialogue data collection in the Waxholm project. *STL-QPSR 1/1995, 50-73.*
3. Bilange, E. (1991). A task independent oral dialogue model. In *Fifth Conference of the European Chapter of the Association for Computational Linguistics (E-ACL'91).*
4. Clark, H. H. (1985). Language use and language users. In Lindzey, G. and Aronson, E., editors, *The Handbook of Social Psychology (3rd edition).* Erlbaum.
5. Clark, H. H. and Carlson, T. (1981). Context for comprehension. In Long, J. and Baddeley, A., editors, *Attention and Performance IX.* Erlbaum.
6. Clark, H. H. and Marshall, C. (1981). Definite reference and mutual knowledge. In Joshi, A., Webber, B., and Sag, I., editors, *Elements of Disourse Understanding.* Cambridge University Press.
7. Cohen, P. R. (1984a). The pragmatics of referring and the modality of communication. *Computational Linguistics,* 10:97–146.
8. Cohen, R. (1984b). A computational theory of the function of clue words in argument understanding. In *COLING'84,* Stanford, CA.
9. Dahlbäck, N. (1991). *Representations of Discourse.* PhD thesis, Linköping University, Sweden.
10. Dahlbäck, N. (1992). Pronoun usage in nli-dialogues: A wizard of oz study. In *Papers from the third Nordic Conference of Text Comprehension in Man and Machine.*
11. Dahlbäck, Nils (1995) Kinds of Agents and Types of Dialogues In J. A. Andernach and S. P. van de Burgt and G. F. van der Hoeven (editors), *Proceedings of the Ninth Twente Workshop on Language Technology, TWLT 9.* Enschede.
12. Dahlbäck, N. and Jönsson, A. (1989). Empirical studies of discourse representations for natural language interfaces. *Proceedings of the 4th Conference of the European Chapter of the Association for Computational Linguistics,* Manchester, England.
13. Dahlbäck, N. and Jönsson, A. (1992). An empirically based computationally tractable dialogue model. In *Proceedings of the 14th Annual Conference of the Cgnitive Science Society (CogSci'92).*
14. Fraurud, K. (1988). Pronoun resolution in unrestricted text. *Nordic Journal of Linguistics,* 11:47–68.
15. Grishman, R. and Kittredge, R. (1986). *Analyzing Language in Restricted Domains.* Erlbaum.
16. Grosz, B. J. (1977). *The Representation and Uses of Focus in Dialogue.* PhD thesis, University of California, Berkeley.
17. Guindon, R., Sladky, P., Brunner, H, and Conner (1986) The structure of user-advior dialogues: Is there a method in their madness? *Proceedings of the 24th Conference of the Association for Computational Linguistics,* 224-230.

18. Guindon, R., Shuldberg, K., and Connor, J. (1987) Grammatical and ungrammatical structures in user-advisor dialogues: Evidence for sufficience of restricted languages in natural language interfaces to advisory systems. *Proceedings of the 25th Conference of the Association for Computational Linguistics.*

19. Guindon, R. (1988). A multidisciplinary perspective on dialogue structure in user-advisory dialgues. In Guindon, R., editor, *Cognitive Science and Its Application for Human-Computer Interaction.* Lawrence Erlbaum Publishers.

20. Jönsson, A. (1993). *Dialogue Management for Natural Language Interfaces.* PhD thesis, Linköping University.

21. Jönsson, Arne (1996) A Model for Dialogue Management for Human Computer Interaction in *Proceedings of ISSD'96, Philadelphia, 69-72, 1996.*

22. Kennedy, A., Wilkes, A., Elder, L., and Murray, W. (1988). Dialogue with machines. *Cognition*, 30:73–105.

23. Lappin, S. and Leass, H. J. (1994). An algorithm for pronominal anaphora resolution. *Computational Linguistics*, 20(4):535–561.

24. Linell, P. (1982) *The Written Language Bias in Linguistics* Studies in Communication 2 (SIC 2). Department of Communication Studies, Linköping University.

25. Malhotra, A. (1975) Knowledge-based English language systems for management support: an analysis of requirements. *Proceedings of IJCAI'75*

26. Pattabhiraman, T. (1994). Review of "user modeling in text generation" by Cécile l. Paris. *Computational Linguistics*, 20:318–321.

27. Rachel Reichman (1985). *Getting Computers to Talk Like You and Me* MIT Press, Cambridge, MA

28. Reilly, R. (1987a). Ill-formedness and miscommunication in person-machine dialogue *Information and Software Technology*, 29:69-74

29. Reilly, R. (1987b) Communication failure in dialogue: Implications for natural language understanding. Paper presented at the seminar *Recent developments and applications of natural language understanding*, London, Dec 8-10.

30. Rubin, A. (1980). A theoretical taxonomy of the differences between oral and written language. In Spiro, R. J., Bruce, B. B., and Brewer, W. F., editors, *Theoretical Issues in Reading Comprehension.* Erlbaum.

31. Stubbs, M. (1983). *Discourse Analysis* Oxford: Blackwell.

32. Volosinov, V.N. (1973) *Marxism and the Philosophy of Language.* New York: Seminar Press.

Using an Interpretation System - Some Observations in Hidden Operator Simulations of 'VERBMOBIL'

Detlev Krause

Universität Tübingen
Fakultät für Informatik
Arbeitsbereich Programmierung
Sand 13, D 72076 Tübingen
email: <krause@informatik.uni-tuebingen.de>

Abstract. The international research project "VERBMOBIL" aims at developing a machine interpretation system capable of interpreting dialogues. This paper focuses on hidden operator simulations, conducted during the first four years of the project. After describing the experimental design, some selected results about factors of acceptance are reported. The most important factors concern the functional elements of an interpretation system, speed on the one hand and completeness, correctness and style on the other. The style of the interpretation seems to be less important. From the perspective of users a machine interpreter should have some additional options but should not have a strong impact on the dialogue itself. Another issue refers to the behaviour of test persons in a simulated Human-Computer-Human Interaction (HCHI). This paper discusses some examples from the study and summarises changes in the human-human interaction as well as interaction between humans and computers. It focuses on obstructions in conversation and provides an indication of possible communication problems. Finally, the results from analysing the typical characteristics of a dialogue enable us to suggest outlines for improving the system.

Social Research for VERBMOBIL

The VERBMOBIL project started in 1993 and is funded by the German Federal Ministry of Education, Science, Research and Technology. The idea was to construct a mobile interpretation system for German and Japanese business managers who have some passive knowledge of English but would like to switch from English to their mother tongues occasionally. In 1996 a prototype was presented that translates German phrases and a small selection of Japanese phrases into English. The domain of the system is currently restricted to appointment scheduling. However, from 1997 on, a system will be realised which will have a wider domain and will include more languages. This will be made available through computer networks, by using mobile telephones, and in media conferences.

The Institute of Sociology at the University of Hamburg was invited to do the accompanying social research for VERBMOBIL. Sociological Research concentrated on three issues; user expectations, technology assessment, and acceptance research. The main method for this last issue was a hidden operator simulation of VERBMOBIL, adapted to the aforementioned scenario of a face-to-face dialogue situation including two speakers and the machine interpreter.

Hidden Operator Simulation (WOZ-study)

Our main purpose was to determine the factors which affect acceptance of the system. In particular, we were intested in what users expect from a machine interpreter in terms of its speed, correctness, completeness, and style. As we could not take advantage of any early version of VERBMOBIL, we developed a hidden operator simulation, better known as a Wizard-of-Oz (WOZ) Study. According to a typical WOZ-Study [1], the wizard represents the system. However, in this case, we could only offer an idea of the future system of VERBMOBIL.

Unfortunately, we had to decide between either introducing a very artificial experimental design with unchanged conditions or a less restricted experimental design open for different scenarios. While the first guarantees a high amount of comparability the second is adaptable to a variety of reseach questions. We decided on the latter, because exploratory research seems to be appropriate as long as there is only little knowledge about how a machine interpreter affects a face-to-face dialogue.

Experimental Design

We used three rooms for the experiments, a reception room, a room for the experiment itself ('dialogue room') and a control room, where the experimenter and the human interpreters listened to the dialogue and provided the interpretation. This interpretation was done by entering the interpreted terms into a personal computer. The basic technical equipment in the dialogue room consisted of

- a 'black box' labelled 'VERBMOBIL'
- special buttons, later mouse buttons, for both speakers to activate the interpretation
- one or two laptops for displaying the interpretation and the system messages for the participants
- loudspeakers for the acoustic output
- tape recorders, sometimes VCRs
- latest version of a common text-to-speech system for English as target language.

Naturally, the technical equipment was improved continuously but did not change fundamentally. The participating individuals were not told about the hidden operator until after the experiment was finished.

Dialogue Scenarios

In contrast to the equipment, the dialogue scenarios were modified several times. The system designers thought it would be easy to simulate the scenarios used for the first versions of VERBMOBIL. However, these versions were restricted to appointment scheduling, and even with a small knowledge of English, people are able to make appointments without any interpretation support. Moreover, they will switch to their mother tongues only when a language problem occurs, which often does not belong to the restricted dictionary of VERBMOBIL. Thus, it proved to be useless to simulate the 'true' VERBMOBIL scenario.

Figure 1 provides a basic overview of the elements of our study. We focused on scenarios related to use situations that can be expected in the future. Therefore, we selected participants who belong to potential groups of users.

It has to be pointed out that these scenarios are by no means the only possible or even the best ones. Every methodological decision is based on intensive discussions about the best way to answer selected research questions. Unfortunately, it is impossible to illustrate this in detail here.

As with the methodological decision making process, the results of the experiment cannot be illustrated in detail, given the constraints of this paper. Therefore, the following presentation of 'results' should be seen as a selection of some arguments and observations which hopefully will support the discussion of dialogue phenomena in a Human-Computer-Human Interaction (HCHI). It is not my intention at all to give final results about the whole nature of HCHI.

Results

1. Code Switching:

VERBMOBIL is designed as an interpretation system that takes advantage of 'code switching.' People initiate the system on demand.

The code switching dialogues in our study are characterised by the difficulties of participants in switching between two or sometimes even three languages. They often become confused, make an English input, or tend to ignore the interpreter completely. Sometimes they wait for an interpretation of an English phrase they do not understand, not realizing that there is no corresponding feature in the simulated machine.

Scenario	Research Interest	Methodical Information
winter 1993 (several students participated)		
some appointments; some free conversations	pre-test, technical means of the simulation	Spanish, Portugheese and German as languages; group discussion; observations
spring 1994 (21 employees or self-employed persons)		
an appointment followed by a free conversation about the participant's work life	exploring main factors of acceptance; user interests	instructed dialogue partner (Turkish, Italian); no code switching; private calendars; computer aided interview (CAI) before, interview after the dialogue
winter 1994 (10 employees or self-employed persons)		
making several appointments for a working week the following year; instructed dialogue partner is advised to refuse or accept certain suggestions made by participants	introduce system messages; stick to the VERMOBIL scenario	instructed dialogue partner (English); using an artificial ('VERBMOBIL'-) calendar; code switching allowed; interview before and after the dialogue
summer 1995 (38 test persons, mostly employees or self employed persons)		
the participant is invited to a talk show in Milan; he is to choose between two ways to go there, is to decide between different leisure offers, and is to determine an appropriate salary	explore the scenario of travel planning and a negotiation dialogue; test the main features of VERBMOBIL (correctness, completeness, style and speed)	instructed dialogue partner (Italian); using a description of the two choices to travel to Milan (by train or by plane); no code switching; accoustic separation of interpretations (male voice) and system messages (female voice); private calendars; CAI before and interview after the dialogue
summer 1996 (6 German and 6 Japanese students)		
slightly different scenarios for each of 3 dialogues: invitation for a dinner; a cultural meeting in the evening; planning a day in Hamburg	explore effects of adaptation by participants using VERBMOBIL repeatedly; explore Japanese attitudes towards VERBMOBIL	only one of 3 dialogues with an instructed dialogue partner (Spanish - German; Spanisch - Japanese); no code switching; no literal instructions; form for completion after every dialogue, group discussion

Fig.1. Scenarios of Hidden Operator Simulations

2. Number of trials:

When system designers asked us to test reactions on system messages, interpretation mistakes, and refused interpretations they thought that test persons would be willing to try again up to four times. Our experience showed, that participants tried two times at the most, and the average participant repeated the same input only once.

3. Relevance of speed and of interpretation quality:

Another research question to explore was the relevance of speed compared with the relevance of quality of interpretation.

Of course, even a professional interpreter needs some time to do her job when asked to type it into a computer, depending on the length and complexity of the phrase. Students or experimenters were interpreting the dialogues in most of the cases. In nearly every experiment we had some English native speakers to interpret or support interpreters but it was impossible to control time and quality of the output in a measurable way. In fact, there are many ways to measure quality of interpretation. It should be expected that potential users of a machine interpreter use other criteria than, for example, linguistic experts.

The following table illustrates the time consumed for the different operations within the hidden operator dialogue:

average length of a dialogue	740 sec.
number of turns of the instructed dialogue partner (Italian) total speaking time (Italian)	13 101 sec.
number of turns of the test person (German) total speaking time (German)	15 83 sec.
average duration of an Italian turn	7.6 sec.
average duration of a German turn	5.4 sec.
total waiting time for the output of an interpr. Italian turn	141 sec.
total waiting time for the output of an interpr. German turn	128 sec.
number of system messages and dismissals	2.8
average waiting period due to a dismissal (counted from the beginning of an input up to the output of the next turn)	52 sec.
average waiting period for the output (including waiting period due to dismissals)	10.7 sec.

Fig.2. Average figures of the 38 dialogues done in summer 1995

There is a strong deviation in most of the figures. For example, the shortest dialogue lasted 7.24, the longest 22.02 minutes.

As the 38 test persons here, nearly all other test persons were not content with the response time while they criticised the quality of the interpretation less. When asked about their main impressions, a complaint about waiting time was most frequently part of the answer. Only when asked to rank the relevance of speed, correctness, completeness, and style of the interpretation did test persons tend to focus on correctness or completeness as figure 3 shows.

The table could easily be misinterpreted, because the terms correctness and completeness are quite ambigous. Some people do not see any difference between these terms. Others evaluate an interpretation already as complete if the relevant information is transferred. One should also bear in mind that demands of correctness and completeness differ according to the relevance of a conversation.

	completeness		correctness		speed		style	
order of priority	1995	1996	1995	1996	1995	1996	1995	1996
1	48	11	37	47	11	33	5	8
2	26	50	29	25	34	17	11	8
3	21	28	26	20	29	25	24	28
4	5	11	8	8	26	25	60	56

Fig.3. Ranking for different features of an interpretation system (in %)

The table demonstrates that people demand correctness and completeness of interpretation. On the other hand, speed is also a highly relevant factor of acceptance becoming even more important when people use the system several times. Only a minority chose style as the most important factor of acceptance.

4. System features:

Concerning the system features we observed that test persons often expect a variety of functions. They do not know about the difficulties of word recognition or disambiguation. When they experience a system translating something successfully, they try to use it in various ways or even communicate with the system. Here is one typical example:

> German test person: No, I don't mean the office here. That's a mistake in the interpretation. I say I want to the mayor's house for/-
> American student: The mayor's house?
> German: Yes, and when you get married you have to go to the Standesamt in German we say and what/- <initiates the interpretation by pressing a button> Was heißt Standesamt? <ends pressing>
> Interpretation: What does registry office mean?
> German: What? That's right.
> American: Oh.
> German: Registry office.

In this example the test person tries to use VERBMOBIL as an acoustic dictionary. As long as such a feature is not integrated into the system, the machine will translate the whole enquiry, as illustrated above, which could cause a comprehension problem for the dialogue partner if she misinterprets a system enquiry as a question she is to answer.

The display of the interpretation seems to be more important than the acoustic output. In summer 1995, when asked their prefered modus of output, people said:

Preference (summer 1995)	N (%)
display	17 (45 %)
acoustic output	3 (8 %)
both are important	17 (45 %)
alternative answer	1 (2 %)

Fig.4. Preferred Modus of Output

Test persons normally had little or no problems to initiate the interpretation by using a button or a mouse. In order to support them, from 1995 on a corresponding German status information was presented on the screen when using the mouse ('VERBMOBIL listens. Please hold until your phrase is finished.').

Some test persons suggested the idea of acoustic initiation. Depending on personal computer experience test persons also demanded different options such as the choice of receiving acoustic and/or literal output, on-line help in the case of unknown terms, a keyboard instead of a microphone, a display of alternative versions of interpretation in advance, data storage, and other additional functions a machine interpreter could provide.

In contrast, they do not want the dialogue to be interrupted by the system. It should be working in the background without any impact on the dialogue. Only in the case of problems do they want to know why the system fails to interpret.

A majority of test persons would prefer an interpretation into their mother tongues rather than into English, as is the case in the VERBMOBIL scenario. On the other hand, only a minority would like to give up the opportunity to control the interpretation by receiving an additional output in a third language.

5. Factors of acceptance in general:

It is not relevant to discuss the test person's statements here in detail because there are many different applications for an interpretation system. Therefore, this would make sense only if we had the chance to discuss concrete options for a concrete purpose. On a general level, we may summarise the test persons' suggestions about an interpretation system as follows:

Computational means for conversation will be accepted for active use as long as they allow for immediate and convenient support. As people expect machine interpreters to rationalise communication, they will use them mainly for purposes of information transfer, including common standards of 'everyday speech'. Therefore, it would be beneficial to provide a wide range of applications according to different purposes. The same applies for different kinds of users. For instance, business managers, who use a system only infrequently, may ask for a simple and modest setup, including higher demands on hardware design. Software engineers, who are familiar with different interfaces, may ask for more sophisticated functions and special terminology allowing technical discussions. [2]

Actions and reactions of users in HCHI

This chapter is about user behaviour in HCHI as we experienced it in our study. Again, it is impossible to present all available results in this paper. It is even more difficult to prove every single argument by examples from our language data. Therefore, the following statements should be used as a stimulous for thought about the characteristics of a HCHI and their relevance for constructing a machine interpreter.

Compared with a dialogue that is interpreted by a human, test persons change their dialogue behaviour in HCHI. Despite of individual differences, there are tendencies to use shorter expressions, to speak slower and louder, to pronounce clearer and to reduce the amount of information. Many people need more time to think before they start to say something.

Success of interpretation has a strong impact on the participants' behaviour. In the beginning, many people are nervous. Depending on place and number of system messages, especially dismissals, test persons feel encouraged to either maintain an obviously successful way of expressing themselves or to change an unsuccessful way. The same phenomenon counts for long waiting periods. During waiting periods a lot of test persons often have 'meta-dialogues' with their dialogue partners - usually in a whisper, as if they do not dare to disturb the interpretation process of the machine.

Here are some examples to illustrate such user behaviour:

German test person: For three hours would be nice. Now we have the 9th of November. Können wir uns am/- pardon <initiates the interpretation by pressing a button> können wir uns am Donnerstag treffen. <ends pressing; 8 seconds waiting period until interpretation, test person whispers:> It would be nice if we could meet on Thursday.
Interpretation: Can we meet on Thursday ?

Here, the test person interprets his German phrase himself after apologising for an input mistake („pardon") because he forgot to press the activating button. Furthermore, he uses a more polite form when he interpretes his phrase himself („It would be nice").

The next example demonstrates a reduction:

German test person: Okay, am Freitag vorm Eingang des Studio Hamburg.
Interpretation (system message): Wortschatzproblem: Bitte verwenden Sie andere Begriffe. <Please use other words.>
German test person: Okay.

Here, the test person refrains from repeating the whole local description that had been given by the Italian dialogue partner previously.

The next example is based on a phonologically similar sound between the terms 'Messe' (fair) and 'Messer' (knife), that the wizard misinterprets in order to simulate a recognition error. Additionally, because the test person makes an English input instead of the expected German, she gets a dismissal.

German test person: Because I have a, ja <initiates the interpretation by pressing a button> Messe <ends pressing; 3 seconds waiting period> <lower tone> fair, ah ja.
Interpretation: Knife
German: <laughter> He doesn't <P> hear correct. <ahm> ah ja/- I don't asked at a knife , I asked at a/-
American student: Maybe - say the whole sentence in German.
German: What? That's right. A fair. <initiates the interpretation by pressing a button> I have to go to the fair. <ends pressing; 3 seconds waiting period>
Interpretation (system message): Bitte wiederholen Sie Ihre Äußerung. Es ist ein Fehler in der semantischen Verarbeitung aufgetreten. <Please repeat. A mistake occured in the semantic process.>
German: Ah ja. <initiates the interpretation by pressing a button> Ich muß zur Messe gehen. <ends pressing; 9 seconds waiting period>
Interpretation: I have to go to a fair.
American: Okay, good. Your English was well, okay.

It is obvious here that the code-switching situation is difficult to handle. The test person gives an English input and the wizard refuses to interpret from English to English. Because the test person did this several times in this dialogue, she remembers to correct herself by repeating the English sentence in German - disregarding the fact that her dialogue partner knew what she wanted to say from the very beginning of this sequence. Of course, this complicates the structure of the dialogue despite the fact that it does not increase the linguistic complexitiy of a turn. Without the wizard's mistake combined with the test person's wrong input, the information concerning the fair could have been delivered earlier.

The figures for the experiments conducted in Summer 1995 illustrate the active and passive shares of a WOZ-dialogue:

Here, the average part of active input is only 11 % for the test persons and 14 % for the Italian dialogue partners. In contrast, 32 % of the time is used for interpreting output and for reception until the next turn begins. 19 % is waiting time for the German turns to be interpreted, and 17 % for the Italian. 7 % of the time is used for clarification between human and machine.

I could present an impressive number of other misunderstandings or problems which emerged during our simulation sessions. The number would certainly increase without a human 'wizard' who is normally, however not in the last example, trying to limit communication problems in order to continue the dialogue. Nevertheless, communication problems characterise our simulated dialogues and will probably characterise 'real' HCHI in future. It is not necessarily the system that produces them, but it seems a crucial point that a system is capable of dealing with them.

The following list summarises some of the main phenomena observed in the simulation sessions:

SPEAKER	LISTENER
⇒ adaptation / controlled input	⇒ increased time for reception
⇒ shift of word and phonem stress	⇒ reinsurance („If I understand you correctly...")
⇒ strong reduction	
⇒ break offs	⇒ ignoring mistakes or problems
⇒ change of subject	
⇒ inconsistent reaction to system messages	⇒ trying to communicate with the speaker directly
⇒ talk to the machine as if it were a bad listener	

⇒ 'meta-dialogues' (about the machine or the situation)
⇒ acoustic signs for displeasure (moan)
⇒ shift of the communication channel (communicate **without** the machine) in case of language problems

⇒ request for a clarification dialogue with the machine (using VERBMOBIL like a language expert)

Fig.5. Selection of typical phenomena in hidden operator simulations (occurring mostly in cases of obstruction)

I would suggest that a machine interpreter has to be prepared to react to the aforementioned problems of HCHI because it functions as a channel for of an information process, transferring signs from a speaker to a listener. According to an ideal model of communication, we may generalise potential communication problems for each part of the HCHI as follows:
The speaker will always have a sending problem because of:

- insufficient use of technical components (here especially the use of the activating mouse button)
- acoustically insufficient input (volume too high or too low)
- insufficient input because of syntactic reasons (complicated structures, incomplete sentences, ambiguos syntactic forms)
- failure of H-C interaction in the case of clarification needs (e.g., if the machine cannot identify speech acts addressing the system)
- failure of H-H interaction because of insufficient self-correction, unnoticed breaks, insufficient information about wrong input, and other reasons.

On the listener's side we find problems of perception:

- insufficient use of technical components (concerning the display or the acoustic output)

- insufficient decoding of the interpretation (on the syntactic, semantic or pragmatic level)
- a speaker-to-computer turn is mixed up with a speaker-to-listener turn (e.g.: „Please repeat this sentence." could be an invitation for the system or for the listener concerning her last turn)
- a computer-to-speaker turn is mixed up with a computer-to-listener turn or with an interpretation (e.g.: „Please repeat your sentence." as a system message could effect in a repetition of the listener's or the speaker's last turn)

It will be easier for a system to detect problems of sending than of reception, because the system has to deal with any input no matter how incomplete it is. In contrast, when a listener does not understand the output, a system is dependent on a feedback about this problem. A speaker will find more indications that a listener has problems understanding him. From our experiments we know that many test persons, particularly those who have little knowledge of English, need quite a lot of time to think about an output.

If one of these problems occurs, a turn-taking decision becomes necessary. In the case of a sending problem, a system can ask the speaker to reformulate her phrase. In case of a reception problem, either the speaker, the listener or the system can find a solution. The speaker may try to make the input simpler, the system may function like a dictionary, or the listener may try to infer what could have been said. Of course, the latter is the normal case, at least in our experience (The reason for this behaviour is probably that a listener prefers not to admit her reception problem. It could disclose her insufficient competence in foreign language speaking.)

The problems caused by VERBMOBIL can neither be categorised as exclusively sending nor as exclusively receiving problems. In the following we list the most important ones:

- insufficient interpretation (wrong, only partly translated, missing interpretation)
- interpretation of something that has not been said
- missing reaction concerning phenomena like hesitation, interruption, repair (especially a sudden desire of turn-taking either by speaker or by listener)
- no detection of technical mistakes (e.g. if a listener presses the button on error)
- no detection of input problems
- interruption of a turn (like a machine, the 'wizard' does not care about any unexpected speech acts and transfers each input as soon as she possible; this is getting difficult if a speaker wants to 'erase' an input or if a listener already understands the speaker directly).

These problems have different effects on the dialogue. For example, if VERBMOBIL interrupts a turn it merely disturbs, whereas, if it is not able to translate, it prevents communication.

Furthermore, we observed psychological strains in participants when they where faced with a communication problem. We detected signs of impatience, stress, and exhaustion. In addition, people sometimes feel helpless, especially when they have difficulties understanding an English expression.

System Design and Dialogue Patterns

At this point, it becomes even more difficult to choose between a system design interfering the face-to-face dialogue as little as possible and another providing solutions for the communication problems mentioned above. An internal model of potential obstructions would certainly help to detect and articulate communication problems. We recommend to take the following fundamental characteristics of a dialogue into account:

- 'flow of communication'
- immediate response
- spontaneity
- structural elements (adjacency pairs [3], start sequence, end sequence)
- hesitations, interruptions, corrections.

System designers who built the prototype of VERBMOBIL would probablably not agree here. Their ideas of a clarification subroutine include the following elements, designed for the prototype of VERBMOBIL:

- rejection of impossible appointment suggestions (e.g., the system would correct a speaker by saying: 'There is no 30th of February.')
- a spelling modus for names
- a yes/no feature for ambigous terms (e.g.: 'Did you say the 13th week?')

The first point is based on the somewhat naive idea that computers should be an intelligent, teacher-like control system of human actions and thoughts. Including this function would result in a change in quality of the machine interpreter. It would then not only be a medium but a 'partner', who knows about the content of the dialogue.

Introducing a spelling modus has been tested in the Japanese-German dialogues of 1996, however, only on the first day because the Japanese students had big problems to spell German names of streets or places. Additionally, the German students said it was very annoying to spell names during a dialogue.

Including a yes/no feature is based on the idea of evoking a clarification between speaker and system, initiated by the system. There is no possibility either for the speaker or the listener to initiate a clarification dialogue by themselves.

All three ideas discussed above focus on a machine's interpretation problem. However, they neglect the options and characteristics of a face-to-face dialogue. One of the most popular ways our test persons chose to overcome a communication problem was to ignore the interpreting facility. Instead, they used other options for transfering a message, i.e. gestures, interpreting the message by oneself, changing the content or the speech act and others. In that respect, they did take advantage of the nature of a dialogue to allow for spontaneous and immediate actions and reactions.

Accordingly, a system should guarantee that these elements of a dialogue can be included. It is also possible, that the inclusion could improve the system itself, as would be the case for corrections, break-offs, turn taking actions, or alterations during the course of a dialogue. Unlike an interpreter, a system does not share the social background of the dialogue partners. While she deals with the physical, emotional and psychological factors of a situation and is able to react immediately to spontaneous communicative events, a system provides only standardised devices to correct, to cancel, or to elicit further (e.g. lexical or linguistic) information. It always performs identically, i.e. there is no adaptation to either changing environmental cirumstances or a changing atmosphere.

A user-friendly construction of a clarification dialogue between system and speaker or listener requires a detailed description of possible communication problems which might emerge in a dialogue. Because dialogues are different, communication problems vary too. Hence, it is advisable to look for characteristics of special kinds of dialogues in order to create help systems adapted to the needs of individual users. [4]

Most of the test persons were looking forward to a device that offers instant interpretation. Although a majority showed a big interest in such a device it was only under the condition that it would serve the purpose they identified as frequentely occurring during their work. However, many of the respondents stated that they would not purchase it for private use only. Ideally, the next step to find solutions for the problems stated above would be an application of an interpreting system constructed according to real needs. As a scientifically oriented project, VERBMOBIL is still far away from this point.

Conclusion

With our study we tried to identify some general directions and shortfalls to be used in future research and design of computer based interpreting systems. We analysed general acceptance of such systems and identified patterns of usage.

Obviously the research we conducted could only serve exploratory purposes. However, for a system to be designed in future in accordance with practical needs our results should be beneficial. Only research that is accompanying the introduction of real word applications that do take the concrete usage situation into account would lead to more definite and applicable results supported by empirical findings.

Finally, on a personal note, I would like to point out that two differing perspectives seem to be the reason for some typical misunderstandings. One is the sociological view on HCHI focusing on changes of more technical mediated communication which can be expected in the future. A differing perspective is based on the technological interest and exclusively focussing on a narrowly defined task to enable technical mediation.

One of these misunderstandings refers to the idea that a machine interpreter should interpret like a human. Hence, system designers try to systematically model and rebuild the patterns a human being follows while interpreting. My approach, on the other hand, is focussing on the identification of principal differences between machine and human interpreting. Hence, the analysis of possible changes in expectations people have when entering a HCHI constitutes a main area of interest. A new instrument rarely substitutes an action but normally only alters it.

The examples from the dialogues outlined above support this argument. In my view, there is a new quality of machine interpreted dialogues that can be characterised by a dominance of technology, the typical patterns of behaviour displayed by the respondents trying to adapt to the input requirements, and indications of stress. People do not expect a system to replace all human abilities. They are prepared to find a short but effective way of communication.

A further misunderstanding is due to our differentiation between HCHI and Human-Computer Interaction (HCI). In HCI a user can perform a particular action, e.g., information retrieval, only by using the interface of the application. In contrast, in HCHI the system is only a device that serves the purpose to support an interaction with another subject. This determines the social quality of an interaction. Hence, a machine interpreter is comparable to a telephone rather than an information retrieval system.

However, this should not be used as an argument against the application of results gained in research on HCI, nor does it mean that widely used methods of userfriendly system designs [5] should not be applicated. My approach describes and understands machine interpreting as a social phenomenon, not only as a technological challenge. Inadvertently, functions and effects of a machine interpreter emerge, which may result in unforeseen system requirements.

References

[1] For a summary about hidden operator simulation, see: Fraser, N.; Gilbert, G.: *Simulating speech systems.* in: Computer Speech and Language (1991) 5, 81-99

[2] Important is also the differentiation between naive, expert and casual users, see for instance in: Paetau, M.; *Mensch-Maschine-Kommunikation. Software, Gestaltungspotentiale, Sozialverträglichkeit.* Ffm., New York 1990

[3] There are a lot of elaborated approaches concerning the design of a system for certain use groups and use cases. In order to suggest dialogue features for H-C clarification we followed: Jacobson, I. et al.: *Object-Oriented Software Engineering. A Use Case Driven Approach.* Workingham, et al. 1992

[4] This could be done according to McIlvenny's suggestions about the potential support of social research: "In the long term, understanding and decreasing the asymmetry between human and machine by incorporating interactivity, presence, local adaption, and repair will be a major development. Much can be gained from studying the:

- Micro-dynamics of intra-turn action: mutual coordination of moments of presence.
- Dynamics of inter-turn activity: constitution of a specific activity.
- Macro-dynamics of activity: routine engagements over time and space." (105)

McIlvenny, P.: *Communicative Action and Computers. Re-embodying Conversation Analysis?* in: Frohlich, D.; Gilbert, N.; Luff, P. (Eds.): *Computers and Conversation.* New York, 1990, pp. 91-132

[5] The concept of adjacency pairs structuring dialogues, is a central topic for conversation analysis. For a summary see: Atkinson, J.M.; Heritage, J.C.: *Structures of social action: Studies in Conversation Analysis.* Cambridge 1984

Classification of Public Transport Information Dialogues using an Information Based Coding Scheme*

R.J. van Vark, J.P.M. de Vreught, and L.J.M. Rothkrantz

Delft University of Technology, Faculty of Technical Mathematics and Informatics,
Zuidplantsoen 4, 2628 BZ Delft, The Netherlands, alparon@kgs.twi.tudelft.nl

Abstract. The goal of a recently started research project of Openbaar
Vervoer Reisinformatie (OVR) is to develop a system to automate part of
its dialogues now held by human operators at its call centres. To achieve
a well thought-out design of such a system both an analysis of the current
human-human dialogues and an experiment determining which form of
dialogues would be favoured by humans were started. In this paper we
will emphasize on the analysis.

Over 5000 dialogues were recorded and transliterated. Our coding scheme
based on information components has been derived from this set. The
coding scheme was applied to a sample of nearly 500 dialogues. Not only
are we able to describe the semantics of the utterances in a dialogue, but
we are also able to classify the dialogues with respect to their difficulty
for an automated speech processing system.

1 Introduction

Openbaar Vervoer Reisinformatie (OVR), a Dutch provider of information concerning national public transport, has recently started an investigation of the usability of Automated Speech Processing (ASP) in their company. In the future, part of their humanly operated information services will be automated using ASP. OVR is involved in European projects on the retrieval of information concerning public transport like MAIS, RAILTEL, and ARISE.

Besides developing a demonstrator for a subset of its information dialogues, the current research project covers the examination of a large corpus of over 5000 dialogues recorded at the OVR call centers. Examining such a large corpus serves a number of purposes:

- developing a standard for characterising information dialogues and services,
- designing and implementing a dialogue manager to manage information transactions,
- examining user expectations concerning such information services and automated versions thereof,

---

* This research was funded by Openbaar Vervoer Reisinformatie (OVR) and SENTER.

– identifying the inherent structures and statistical relationships of dialogues by empirical methods.

This article is aimed at describing an effective coding scheme to analyse information dialogues and to describe how the level of difficulty of information dialogues can be measured based on an unambiguous coding scheme and a large corpus.

Related work is discussed first, followed by a description of the dialogues examined. In the fourth section, the coding scheme based on these information dialogues is discussed. Results of applying this coding scheme to OVR dialogues will be presented in section 5, followed by a discussion of automated processing of these dialogues. Finally, future work will be described.

2 Related work

The basic assumption of a plan based theory for dialogue modeling [4] is that the linguistic behaviour of agents in information dialogues is goal directed. To reach a particular state agents use a plan which is often a small variation of a standard scenario. The structure of a plan resembles the dialogue structure. Dialogue acts in OVR conversations are also assumed to be part of a plan and the listener is assumed to respond appropriately to this plan and not only to isolated utterances.

Carletta et al. [3] developed a coding scheme for task oriented dialogues existing of three levels. At the highest level dialogues are divided in transactions or subdialogues corresponding to major steps in the participants' plan for completing the task. At the intermediate level conversational games consist of initiations and responses. A conversational game is in fact a sequence of utterances, starting with an initiation and ending with the requested information being transferred or acknowledged. At the lowest level conversational moves are utterances which are initiations or responses named according to their purpose. Our coding scheme has been strongly influenced by their scheme, however, our scheme is both more generic and more detailed.

In the VERBMOBIL project [1] researchers developed a taxonomy of dialogue acts. To model the task oriented dialogues in a large corpus they assumed that such dialogues can be modelled by means of a limited but open set of dialogue acts. They defined 17 dialogue acts and semi-formal rules to assign them to utterances. These assignment rules served as a starting point for the automatic determination of dialogue acts within the semantic evaluation system. From the analysis of an annotated corpus of 200 dialogues they derived a standard model of admissible dialogue act sequences (nowadays they use over 40 dialogue acts and their corpus has risen to round 500 dialogues). We also investigated phase sequences in a similar way (see section 5).

Another approach of modeling dialogues [2,10] is based on the observation that in goal directed dialogues, the information is exchanged in such a way that the dialogue contains one or more topical chains. These are sequences of utterances that all communicate information about the same topic (topic packets).

Successive topics are usually related to each other. Topic transitions in a dialogue are modelled as movements across these topic packets. The local relationship between topics can be used in knowledge based systems in a dialogue manager for ASP systems.

Since the introduction of plan theory researchers criticized a plan based approach. For example Dahlbäck [5] noticed that in advisory dialogues it is possible to infer the non-linguistic intentions behind a specific utterance from knowledge of the general task. In an information retrieval task this inference can be very difficult, however, in order to provide helpful answers this information is not needed frequently. Since the OVR dialogues are strongly goal directed, intentions are not very important. In a recently developed demonstrator for the German railways, intentions are deliberately neglected. This system is strongly goal directed as well [9].

3 Dialogues

The coding scheme described in this paper, is developed using a set containing 5205 dialogues [11]. These dialogues consist of telephone conversations between two human agents. One of these agents is an operator working at a call center, while the other agent is a client calling the call center to gain information.

The context in which these conversations take place is well defined, since all clients want to retrieve information about the services offered by the Dutch public transport system. The piece of information a certain client is looking for can range from simple fare information to more complex information like complete travel schedules between two streets anywhere in the Netherlands using all different kinds of transport modalities.

The goal both agents pursue is quite clear. Clients want to get the information about public transport which they desire and the operator wants to serve the client as good as possible. However, sometimes information is presented which is not of interest to the client or of which the client is already aware of.

duration	number		duration	number	
(sec.)	#	%	(sec.)	#	%
0 - 10	1	0.0	105 - 120	283	5.4
10 - 20	122	2.3	120 - 150	294	5.6
20 - 30	590	11.3	150 - 180	185	3.6
30 - 40	740	14.2	180 - 210	118	2.3
40 - 50	677	13.0	210 - 240	69	1.3
50 - 60	504	9.7	240 - 270	55	1.1
60 - 75	670	12.9	270 - 300	29	0.6
75 - 90	479	9.2	300 - 360	30	0.6
90 - 105	326	6.3	360 - ∞	33	0.6

Fig. 1. Dialogue duration

Two major tasks can be distinguished in all conversations. The first task is to get both parties to understand which information is desired by the client and which information is needed to query the public transport information database. The second task consists of the operator presenting the information inquired by the client. The information is normally broken down in several pieces which makes it easier for the client to memorise or to write down the information. In most conversations both tasks only occur once or twice. However, they may occur more often.

In figure 1 the frequence table is given for all conversations, which have been recorded by the Dutch PTT during spring 1995. Actually, the dialogues were grouped into four categories: train information, bus/tram information, train & bus/tram information, and miscellaneous [11]. The median of the dialogues containing train information only is approximately $\frac{3}{4}$ minute while for all other categories the median lies around $1\frac{1}{4}$ minute.

Before dialogues were coded, they were transliterated: spoken text was transcribed into written text. Transliteration was necessary to gain better insight into the mechanisms underlying the studied dialogues and eventually to derive a coding scheme based on these information dialogues.

```
2:goedemorgen reisinformatie               Gre(G,[],[])
  (good morning travel information)
1:goedemorgen [achternaam] kunt u mij zeggen hoe   Gre(G,[Per([Nam])],[]),
  laat de[uh] bus van Lochem naar Deventer toe gaat Que(Q,[Tt([DaS([Unspecified])])]),
  (good morning [last name] can you tell me what)    RI([TrT([Bus_Tram(0)])]),
  (time the[uh] bus departs from Lochem to)          Loc([DeP([City(1)])]),
  (Deventer)                                         ArP([City(2)])])]),[])
2:hoe laat ongeveer zou u mee willen       Que(S,[Tm([DeT([Unspecified])])],[])
  (how late approximately would you like to go)
1:[uh] rond een uur of negen               Sta(Q,[Tm([DeT([About(3)])])],[])
  ([uh] about nine o'clock)
2:oké                                       Ack(Q,[Ack([Pos])],[])
  (ok)
2:negen uur twee mevrouw buslijn zesenvijftig  Sta(I,[Tm([DeT([Exact(4)])]),
  (nine o two madam bus line fifty-six)          RI([TrT([Bus_Tram(5)])])])],[])
1:oké dank u wel                            Ali(B,[Ann],[]),Bye(B,[Bye([Tha])],[])
  (ok thank you)
2:tot uw dienst                             Bye(B,[Bye([ReG])],[])
  (at your service)
1:ja hoor dag                               Bye(B,[Bye([ReG,Goo])],[])
  (yes fine bye)
2:goedemorgen                               Bye(B,[Bye([Goo])],[])
  (good morning)
```

Fig. 2. Sample dialogue and coding (agent 1 = client, agent 2 = operator, for other abbreviations see next section)

4 Coding Scheme

Goals of the current research project are to identify inherent structures and statistical relationships of dialogues and to develop a taxonomy of information dialogues based on these structures and statistical relations between move and phase transitions. The main point of interest is how information is transferred between agents. We therefore developed a coding scheme based on the coding of information and on dialogue acts used to transfer information (see [13] for a detailed description).

The primary objective of the coding scheme was to analyse the previously mentioned dialogues. This version of the coding scheme was not intended to be used in the dialogue manager of the future ASP system, although the coding scheme to be used by the dialogue manager will be based on this coding scheme. The advantage of a separate coding scheme for analysis is that we could totally focus on the analysis and not to let dialogue manager design details dictate our coding scheme.

4.1 Outline

Dialogues consist of turns which normally alternate between agents. These turns consist of one or more utterances which correspond to grammatical sentences. These utterances are called moves which are coded as comma separated terms.

A move takes place in a certain phase of the conversation which indicates whether the operator or the client has the initiative. This phase is coded as an argument of the move because moves play a more important part in the analysis of the dialogues than phases. Therefore we prefer to tag moves by their phases instead of placing moves hierarchically lower than phases.

Other arguments of moves are symbols used to code information, like Location, Time, etc. Information codes can have layers of subcodes with more detailed information (see figure 3). The final argument of a move is used to code interruptions.

To give an example from figure 2 where the operator said 'nine o two madam bus line fifty-six', this is a statement, Sta, happening in the information phase, I, about the exact departure time, Tm([DeT([Exact(4)])]) (i.e. Time, Departure Time, and Exact), of a bus, RI([TrT([Bus_Tram(5)])]) (i.e. Route Information, Transport Type, and Bus or Tram), and this statement was not interrupted, [].

4.2 Phases

Dialogues can be divided into several phases or subdialogues which accomplish one step in retrieving the desired information. Phases are typically a number of consecutive topic packets [10] used to complete such a step in the dialogue. Examples of subdialogues modelled by phases are to define which information is needed or to present the information to the client.

Phases are important aspects of our coding. A dialogue phase is a good indication of the flow of information at a certain time in the dialogue and it denotes which agent has the initiative. For instance, during the Greeting phase both agents present theirselves and in some case they also give information about their residance ('good morning with [lastname] from Rotterdam'), while in the Query phase the client has the initiative and presents most information. Another important aspect of dialogue phase is its usability as a measure of dialogue complexity. In our coding scheme, six phases can be distinguished.

During the *Greeting (G)* phase at the beginning of the dialogue, both parties exchange greetings and often introduce themselves.

In the *Query (Q)* phase the client has the initiative and tries to pose his question. All parameters which contribute to the question can be altered during this phase. For instance, the client can mention both the arrival time and departure place during one instance of the Query phase. In some dialogues there is no explicit question ('Tomorrow I have to be in Rotterdam'). When the information given by the client is insufficient to find a solution, the operator can elicit further information by posing supplementary questions.

Periods of silence occur during dialogues, indicating that the operator is searching. These *Pauses (P)* can often be found between Query and Information phase and are frequently introduced by the operator ('one moment please').

During the *Information (I)* phase the operator provides the information asked for by the client. The operator who has the initiative, often presents the information in small pieces to make it easier for the client to memorise or to write down the information. Each piece is usually acknowledged by the client.

The reason for having a *Subquery (S)* phase besides the Query phase is to enable detection of inadequate or incomplete information transfer. This makes it possible to develop a more precise dialogue taxonomy (see section 6). Subqueries are phases which vary only a small part of the original query and always indicate a switch of initiative.

Subqueries can originate from a lack of information provided for by the client or from imprecise or wrong information provided for by the operator. Subqueries can therefore both be initiated by the operator and by the client. Operator initiated subqueries are usually supplementary questions to retrieve information from the client which has not been given and which is necessary to extract the information asked for. Client initiated subqueries are strongly related to the original query.

At the end of a dialogue, the *Goodbye (B)* phase is often introduced by an alignment, which is used to check whether both client and operator are ready to end the conversation by exchanging thanks and saying goodbye.

Looking at these phases, in only three of them relevant information is exchanged: Query, Subquery and Information. The other phases serve social goals, because information dialogues have to observe norms and conventions of both agents' cultures.

4.3 Moves

Phases consist of one or more conversational games [3]. These games usually consist of initiations, e.g. questions or statements, followed by responses, like

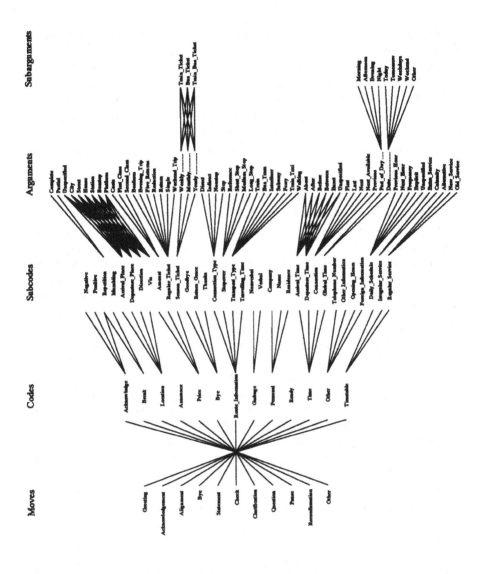

Fig. 3. Outline of Coding Scheme

answers or acknowledgements. Moves are simply initiations and responses classified according to their purposes.

Whenever an agent asks for a piece of information, this is coded as a *Question (Que)*. Implicit questions (e.g. ones that start with 'I like to know ...') are also coded as a question. Information can be introduced as part of a question.

Statements (Sta) are utterances with the purpose of presenting information in response to a question posed earlier in the dialogue or of describing the context of a new question. Typically, this information is not given in an interrogative form.

An *Acknowledgement (Ack)* is a verbal response which minimally shows that the speaker has heard the move to which it responds. It often also demonstrates the acceptance or rejection of the move which is acknowledged.

A *Check (Che)* requests the partner to acknowledge information that the checker has some reason to believe, but is not entirely sure about. Typically, the information to be confirmed is something which the checker believes was meant to be inferred from what the partner has said. For instance, when a client mentions Rotterdam as the destination, the operator usually checks if Rotterdam Central Station was meant as the destination of the journey.

A *Reconfirmation (Rec)* requests the partner to repeat the answer or question given. This occurs in three cases: the person who asks for reconfirmation did not hear the partner's move, the person is aware of some forgotten information, or the person does not believe a given answer.

A *Clarification (Cla)* is an utterance which presents a piece of additional information which was not strictly asked for by the other agent. A clarification can be a reply to a question in which case it does not present the precise information asked for (e.g. the response 'a little bit later is also fine' on the question 'at 9 o'clock?'). A spontaneous clarification is also used to present information in addition to the information given in the previous statement.

An *Alignment (Ali)* verifies the attention and/or agreement of the partner to move to a next phase. The most common type of alignment requests the partner to acknowledge whether the query has been answered to satisfaction or whether the client wishes to drop the query and that they are both ready to move on.

Some moves are used for coding social behaviour occurring during information dialogues. These moves do not serve goals other than confirming to social behaviour normally displayed in such dialogues. In our coding scheme we use the following 'social' moves:

Greeting (Gre) is the move in which an agent greets the other agent and often introduces himself.

The *Bye (Bye)* move describes the social ritual where the client thanks the operator, who on her term expresses that she was glad to help, after which both parties say goodbye.

A *Pause (Pau)* is an indication that the operator is busy processing the query and that during the pause no information is exchanged.

Finally, we also have a code *Other (Oth)* for verbal and nonverbal garbage (like unintelligible utterances and noise).

4.4 Coding of Information

Information codes are used to code what type of information was transferred in every move described above. Coding of information was done hierarchically to facilitate automated processing of coded dialogues. An example is Location, followed by an information subcode, e.g. Departure Place. Arguments to this information code are items like Street or City. In a few cases, arguments have subarguments.

Arguments are sometimes suffixed by numbers indicating some piece of information. For instance, all references in a dialogue to Amsterdam could be coded as 'City(1).' Based on the coded dialogues it is possible to detect when new pieces of information are introduced or when references are made to information introduced earlier.

In this way examining information references is possible without the obstruction of the huge amount of information presented and repeated during a conversation. Translation of the coding back to the original conversation is possible, albeit without the exact words but with the correct information references.

Location (Loc) contains information about locations mentioned in the dialogue. This can be the arrival and departure place, but also intermediate stations or directions. Typical arguments for location are City, Station, etc.

Time (Tm) concerns the general time setting of the query. This includes arrival or departure time and date of travelling, but also relative time settings like next or last connection. Arguments for this code are typical time and date references.

Route Information (RI) covers the type of the journey to be made. This information ranges from Transport Type (e.g. Bus or Train) and Connection Type (e.g. Direct or Nonstop) to Stopovers and Travelling time.

Timetable (Tt) concerns all information producing insight into the general set-up of the timetable belonging to a specific journey. This can be information of a regular timetable (like 'twice every hour') as well as information concerning changes in regular timetables as an effect of irregularities.

Price (Pr) covers the costs of all types of public transport tickets. This covers regular tickets, like Single and Return tickets, as well as Reduction and Season Tickets.

Acknowledge (Ack) is used to code all acknowledgements in response to both questions and statements. Acknowledgements can be either Positive, Negative or Repetitions.

Other (Oth) represents all information which can not be coded by any of the other codes. Typical examples are telephone numbers. Social chit-chat is also coded as Other (the difference with verbal garbage is that social chit-chat is intelligible while verbal garbage is unintelligible).

64

Three codes are used to code control information. *Announce (Ann)* marks information concerning the announcement of a Pause or Alignment. *Break (Bre)* concerns any irrelevant moves during a pause in the conversation which can be utterances to report progress or to break silence. *Ready (Rea)* is used to get the attention of the partner, typically at the end of a pause.

Social information is represented by the following codes. *Personal (Per)* is used to model personal information, like Name and Residence. *Bye (Bye)* is used to code the final moves of a dialogue. Typical arguments are Thanks, Return Grace ('at your service'), and Goodbye. *Garbage (Gar)* is anything irrelevant to the goal of the dialogue.

5 Results

The coding scheme was applied to a sample of 483 dialogues. We have chosen not to code 37 of them since they would not help us designing an ASP system. For instance one dialogue was about an extremely worried parent whose daughter is missing. This conversation will be unlikely occur with an ASP system. Another example is a dialogue entirely in a foreign language. Since our speech system will work with Dutch only, this dialogue was irrelevant for our purpose. Although in theory we could have coded these dialogues, we have chosen to call these 37 dialogues 'uncodable.'

It was to be expected that a larger part of the longer dialogues would be uncodable. This turned out to be true, but the correlation was not as strong as expected as half the uncodable dialogues were shorter than two minutes and a quarter was even shorter than one minute.

The coding process has been performed by 3 men. The first 25 dialogues were identical for all 3 coders. After 10 and 25 dialogues there were review sessions to determine if the coding did not diverge too much. After the first 25 dialogues the remaining dialogues were distributed among the coders. During the coding of those dialogues the coders consulted each other frequently.

5.1 Phase Transitions

We started by examining the phase transitions that took place. The transitions between the phases Q, S, P, and I are eminent, which was to be expected. The core of the conversations (see figure 4) consists of queries, pauses used to search the databases, and by information supply.

As expected a Query phase follows the Greeting phase and just before the Bye phase in most cases information will be given. The other cases are infrequent but nevertheless interesting.

The 5 cases with a QB transition are important since all theses dialogues ended after a question and were not followed by an answer. Further investigations showed that the operator could not help the client. In most cases the client did not know the exact location where he wanted to go. The public transportation information system can cope with exact addresses but not with names of buildings (like conference centres).

The cases of BQ & BS transitions are typically clients interrupting the Bye phase because they realise that they have another question. On the other hand a BI transition typically occurs when an operator interrupts the Bye phase by giving the client extra information that just popped up.

In the 2 cases where there was a GP transition, the clients were not ready yet to formulate their question. They were probably distracted when they were on hold. The case of the PG transition was caused by a change of operator. The two cases of the PB transitions were quite different: in one dialogue the client interrupts the operator's search by saying that he has enough information while in the other dialogue the client took a long time to copy down the information.

	G	Q	S	P	I	B
G	-	444	0	2	0	0
Q	0	-	181	282	267	5
S	0	199	-	55	174	0
P	1	122	69	-	262	2
I	0	75	175	99	-	443
B	0	9	9	0	38	-

Fig. 4. Phase transitions

5.2 Phase Automaton and Profiles

For an ASP system we found the QB, BQ, BS, and the BI transitions of the infrequent cases important enough to set a threshold on 5 dialogues for a transition to be significant. Out of the remaining cases we have constructed an automaton (see figure 5) that can accept the phases that occur in dialogues (including non-changing phases).

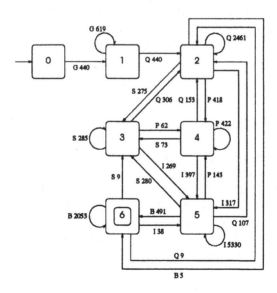

Fig. 5. Phase automaton (initial state = 0, final state = 6 and transitions are labeled by phase and the number of transitions found in the sample.)

Of all dialogues just 6 are rejected: 5 because of an illegal transition (caused by the transitions that did not reach the threshold) and in one dialogue the final state was not reached. Although tempting to believe, the states in the diagram have no direct link with the phases in a dialogue but are due to the automatic construction of the automaton.

When we look at the phase transitions of the dialogues we find that a few phase profiles (i.e. strings describing the phase transitions) occur frequently (see figure 6). Short dialogues occur more frequently and the diversity in short profiles is limited. Although a larger variety in longer profiles can be found, these could easily be clustered into similar cases.

Phase Profiles	#	%
GQIB	51	11.4
GQPIB	35	7.8
GQPQIB	15	3.4
GQSQIB	14	3.1
GQPISIB	12	2.7
GQSQPIB	11	2.5

Fig. 6. Phase profiles occurring more often than 10 times in the 446 coded dialogues

The profiles GQIB, GQPIB, GQPQIB, and GQPISIB all have one commonality: the client supplies all the information needed such that the operator is capable to answer the query without any urge of the operator. In the profiles GQSQIB and GQSQPIB the operator asks the client to supply further information so that the operator can answer the client's question.

The coded dialogues have also been examined at the level of moves, but as space is limited, we refer to [11,13] for a detailed description.

6   Classification

Although the difficulty of a dialogue (with respect to the complexity of an ASP system to handle such dialogues) is proportional to its length, it is also possible to make a second taxonomy of the difficulty of dialogues. One of our objectives is to investigate the potential difficulties for ASP systems occuring in dialogues.

We have defined five types of dialogues. Each of these succesive types needs a more powerful ASP system to recognise the dialogues. These types do not vary with dialogue length or the complexity of sentences. We distinguish the following dialogue types:

1. The client presents all necessary data. In the easiest form this can be regarded as filling in a question form in free order. Once all slots are filled, the operator is able to retrieve the information requested.

2. The operator continues to ask for additonal information until the client has given all necessary data. Often a client has forgotten to fill in all slots and the operator keeps on asking for the missing slots. In many cases the client is not aware what information is necessary for the operator to be able to retrieve an answer from the information system.
3. The operator helps with suggestions. Sometimes the client does not know the exact location where he wants to go to. Often the operator can help out by suggesting a place nearby.
4. There is insufficient data for the operator. The operator cannot help the client with his question and the dialogue stops without a travel advice.
5. Miscellaneous. Actually, this is none of the above. It can also be seen as the type of conversation that would not be processed by a computer (at least at this stage).

With our coding scheme we can recognise these types using phase transitions and the occurence of certain moves (disregarding the Pause phase):

1. These dialogues go directly from the Query phase to the Information phase and they do not belong to type 3. These dialogues do not have a Subquery phase in the Query phase.
2. Just as type 1, but with a Subquery phase in the Query phase.
3. Just as type 2, but the dialogues contain a Check move initiated by the operator in the Query phase and in the Check new information is presented.
4. These dialogues do not have an Information phase before the Bye phase but a Query phase.
5. These were the uncoded dialogues.

Using several tools we were able to establish the type of all coded dialogues (see figure 7). It was a bit surprising to see that the number of conversations with help of the operator was small and that the number of conversations where the client's question remains unanswered was even smaller.

An ASP system that works with question forms could in theory handle all cases of type 1 and 2. This leads to an upper bound of 88% of all conversations that could be processed by form based systems. In practice this upper bound is far too optimistic knowing the current state of art.

The reason for this is that spoken text does not resemble written text. Most spoken sentences can be regarded as a concatena-

Type	#	%
1	134	28
2	291	60
3	17	4
4	5	1
5	37	8

Fig. 7. Dialogue types

tion of short sentence fragments of a relatively easy structure. It is the speaker's hope that the listener will be able to understand the meaning of the spoken sentence. In most cases the entire sentence is not grammatical but the fragments are fairly grammatical [12] (this study was for Japanese). Ambiguity is less a problem in spoken text than it is in written text. To give two examples: subordinate clauses and recursive preposition phrases are far less frequently used in spoken text than in written text.

This leads quite naturally to the need of a parser that is capable of coping with non-grammatical sentences, missing words, and incorrect words: a robust parser (see [6,8]). Many robust parsers first try to parse the sentence in a non-robust way and when this fails they incrementally try to adjust the partial parse tree or sentence in such a way that parsing can continue. Performance deteriorates with the number of adjustments. A sentence with more than a couple of mistakes will hardly be parsable in real time.

The bottleneck in this scheme will be the speech recogniser [7]. Even with a near perfect speech recogniser we cannot process more than $\frac{2}{3}$ of the dialogues (based on a small sample). Dialogues about train information are favourable. The vocabulary for such dialogues is very suited: the number of stations is about two orders smaller than the total amount of proper names (a.o. streets and cities) accessible in the current information system. Also it is quite important that on average these dialogues are much shorter.

7 Future Research

Since we will work on an ASP demonstrator in the long term, we have planned several subgoals along the way. As a first step we have started two independent studies.

The first one, as described in this paper, was to analyse the dialogues between clients and operators using the current system. Although we already have a substantial amount of statistics, we still need more statistics on a more detailed level. To get usable statistics we need to code more dialogues. Currently we manually code the dialogues. We plan to work on a way to do this automatically or semi-automatically (i.e. human assisted).

The second study is an experiment [11] where 409 people were instructed to plan a trip between two places at a certain time. The persons were asked to use the current system and to use an ASP system. The persons were not aware that the ASP system was actually a simulated system operated by a human. The travel instructions given to the participants were made in such a way that they had to formulate their own questions to the ASP system. The simulated system was not the same for all people, we diversified the way the system worked so that we were able to determine which type of human-machine dialogues was appreciated best.

As a second step we evaluated available ASP systems by using our coding scheme and classification. Together with the results obtained from the research described in this paper and the second study, we plan to design a suitable dialogue structure for a dialogue manager that can be used in the ASP demonstrator. Such a structure will make use of a coding scheme devised on the coding scheme described in this paper. Concurrently we will work on the development of a speech recogniser.

In the final step we will try to fit all into a prototype that will show the strengths of the techniques used. If this prototype works satisfactorily, it will be reimplemented more efficiently and in parallel by others such that it can be used by OVR. It is hoped that the ASP system will supplement the current operators such that their workload can be reduced. We hope that the ASP system will take over the easier client questions, while the operators can concentrate on the more interesting questions.

References

1. J. Alexandersson and N. Reithinger. Designing the dialogue component in a speech translation system, A corpus based approach. In *Corpus-based Approaches to Dialogue Modelling*, 9th Twente Workshop on Language Technology, pages 35–43. University of Twente, 1995.
2. H.C. Bunt. Dynamic interpretation and dialogue theory. In *The Structure of Multimodal Dialogue*, volume 2. John Benjamins Publishing Company, Amsterdam, 1995.
3. J. Carletta, A. Isard, S. Isard, J. Kowtko, G. Doherty-Sneddon, and A. Anderson. The coding of dialogue structure in a corpus. In *Corpus-based Approaches to Dialogue Modelling*, 9th Twente Workshop on Language Technology, pages 25–34. University of Twente, 1995.
4. P.R. Cohen. Models of dialogue. In *Cognitive Processing for Vision & Voice*, 4th NEC Research Symposium, pages 181–204. NEC & SIAM, 1994.
5. N. Dahlbäck. Kinds of agents and types of dialogues. In *Corpus-based Approaches to Dialogue Modelling*, 9th Twente Workshop on Language Technology, pages 1–11. University of Twente, 1995.
6. D. van der Ende. Robust parsing: An overview. Technical Report Memoranda Informatica 95-03, University of Twente, Department of Computer Science, 1995.
7. E. Keller, editor. *Fundamentals of Speech Synthesis and Speech Recognition, Basic Concepts, State-of-the-Art and Future Challenges.* John Wiley & Sons, 1994.
8. R. Kiezebrink and J.P.M. de Vreught. An annotated bibliography on robust parsing and related topics. Technical Report CS 96-58, Delft University of Technology, 1996.
9. M. Oerder and H. Aust. A realtime prototype of an automatic inquiry system. In *International Conference on Spoken Language Processing*, pages 703–706, 1994.
10. M.M.M. Rats. *Topic Management in Information Dialogues.* Ph.D. thesis, Tilburg University, 1996.
11. L.J.M. Rothkrantz, R.J. van Vark, J.P.M. de Vreught, J.W.A. van Wees, and H. Koppelaar. Automatische spraakherkenning en OVR-dialogen. Technical report, Delft University of Technology, Knowledge Based Systems, 1996. in Dutch.
12. M. Seligman, J. Hosaka, and H. Singer. "Pause Units" and analysis of spontaneous Japanese dialogues: Preliminary studies. In *Workshop Dialogue Processing in Spoken Language Systems*, European Conference on Artificial Intelligence, 1996.
13. R.J. van Vark, J.P.M. de Vreught, and L.J.M. Rothkrantz. Analysing OVR dialogues, coding scheme 1.0. Technical Report CS 96-137, Delft University of Technology, 1996.

Speech Production in Human-Machine Dialogue: A Natural Language Generation Perspective[*]

Brigitte Grote[1], Eli Hagen[2], Adelheit Stein[3], Elke Teich[4]

[1] Otto-von-Guericke Universität Magdeburg, IIK, Universitätsplatz 2, D-39106 Magdeburg
[2] Deutsche Telekom, FZ131, PO Box 100003, D-64276 Darmstadt
[3] German National Research Center for Information Technology (GMD-IPSI), Dolivostr. 15, D-64293 Darmstadt
[4] University of the Saarland, Department of Applied Linguistics, Translation and Interpretation, PO Box 151150, D-66041 Saarbrücken

Abstract. This article discusses speech production in dialogue from the perspective of natural language generation, focusing on the selection of appropriate intonation. We argue that in order to assign appropriate intonation contours in speech producing systems, it is vital to acknowledge the diversity of functions that intonation fulfills and to account for communicative and immediate contexts as major factors constraining intonation selection. Bringing forward arguments from a functional-linguistically motivated natural language generation architecture, we present a model of *context-to-speech* as an alternative to the traditional *text-to-speech* and *concept-to-speech* approaches.

1 Introduction

Both speech recognition and speech synthesis research in the context of dialogue systems take an analysis perspective on speech. Recognition of spontanuous speech deals with incomplete and incorrect input, which involves syntactic and semantic analysis of the speech signal. In speech synthesis, the major issue is how to best render a written text into speech, i.e., a text is subject to morphological, syntactic, and semantic analysis. Even though synthesis implies generation, combining speech production with natural language generation has not been seriously considered as an alternative to text-to-speech approaches.

In terms of application of speech technology, both academia and industry appear to focus their attention on automatic translation of human-human dialogue in different languages (see e.g., the German VERBMOBIL project) and on telephone interfaces to information systems. We think that these application domains encourage an unbalanced concentration of effort on speech recognition.

[*] Authors appear in alphabetical order.—This work was partially funded by the European Union Programme Copernicus, Project No. 10393 (SPEAK!) under contract with the Darmstadt University of Technology. All authors have been actively involved in the project at various stages, either under employment at the Darmstadt University of Technology or GMD-IPSI.

In both application domains, efforts have been concentrated on analysis and interpretation of the incoming speech signal. Also, speech production in these types of application has mostly been concerned with rendering already existing text into speech, where the text to be synthesized is canned or constructed from predefined templates[5].

In this article we discuss an alternative approach to speech production. It moves away from considering speech production as an analysis task towards a natural language *generation* view, in which the text to be synthesized is automatically generated. The application context is a dialogue interface to an information retrieval system, in which a dialogue is analyzed in terms of speech acts building up a dialogue history that defines the context in which the natural language generation process operates. This approach bears the promise of overcoming some of the problems encountered by strictly analytical text-to-speech systems, such as the problem of assigning appropriate intonation contours. It is hardly possible to extract prosodic features by syntactico-semantic analysis, because the different types of information involved in prosody are not available or cannot be inferred. Since in our approach prosody assignment is an integral part of the generation process, we simply avoid the limitations of automatic analysis. Clearly, speech technologies must solve speech-specific problems, but since the problem domains for written and spoken language technologies overlap, we argue here for an integration of speech synthesis and natural language text generation. On the speech recognition side, a similar train of thought encouraging the integration of natural language understanding and speech understanding has been set forth, for instance, by Black & Campbell (1995) and Hirschberg, Nakatani, & Grosz (1995).

In our work we concentrate, in particular, on the problems of choosing syntactic form and intonation patterns appropriate for a given context. Combining a state of the art dialogue model (Sitter & Stein, 1996) with a state of the art natural language text generator (Bateman & Teich, 1995), we determine the kinds of pragmatic and linguistic knowledge needed to constrain syntactic and intonational choices. From the dialogue model we construct an explicit representation of the dialogue history. We thus introduce an approach that moves away from *text-to-speech* generation towards *context-to-speech* generation, in which both *communicative* and *immediate* contexts are employed for constraining intonation selection.

Our approach was developed in the framework of SPEAK!, a joint collaborative project between the technical universities of Darmstadt and Budapest (cf. Hagen & Stein, 1996; Olaszy, Gordos, & Nemeth, 1992; Teich *et al.*, 1997). The main goal of the project was to develop a generic dialogue-based interface to information retrieval systems with a spoken output channel. An overview of the prototype's system architecture is presented in Figure 1. The SPEAK! prototype offers access to a large full-text database in the domain of art and art history. The input channel is written and direct manipulation, and the user queries are

---

[5] For a representative collection of various types of spoken dialogue systems see Dalsgaard *et al.* (1995) and LuperFoy, Nijholt, & van Zanten (1996).

processed by the INQUERY retrieval engine (see Callan, Croft, & Harding, 1992).
The spoken channel is used to convey meta-communication that complements
the visual output and direct manipulation input. A dialogue history defines the
context in which to generate comments and the text generator produces utterances with intonation markings that are interpreted by the MULTIVOX speech
synthesizer (see Olaszy, Gordos, & Nemeth, 1992).

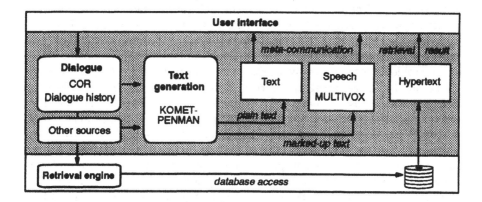

Fig. 1. The SPEAK! architecture

The example dialogue below illustrates a typical interaction sequence between
user (U) and system (S), in which all of the system's utterances are output in
spoken mode. Selected examples of the assignment of intonational features to
the system's utterances are discussed throughout the article.

Example dialogue

S: *Herzlich willkommen in diesem elektronischen Dialogsystem. Mit diesem
System können Sie ...*
[explaining how to use the system, e.g., how to enter and refine queries]
U: enters query terms: "architect" and "Art Nouveau"
S: *Einen Moment bitte, ich durchsuche die Datenbank.*
[telling the user that database is being accessed]
*Ich habe eine Reihe von Artikeln gefunden. Ihre Titel sind unter Ergebnisse
aufgelistet. Wenn Sie einen ansehen wollen, müssen Sie nur ...*
[informing user that many items have been found, displaying the titles
of the articles found, and explaining how to access the full texts]
U: selects "Charles R. Mackintosh" from list,
scans through the biography displayed, and
enters additional search term to refine previous query: "Darmstadt"

S: *Möchten Sie den Artikel über Mackintosh vorher speichern?*
[asking user if document should be saved]
U: clicks on save
S: *Gut, ich behalte den Artikel*
[saving article]
und bearbeite Ihre neue Anfrage. Moment bitte.
[processing the modified query]
*Die neue Ergebnisliste ist da. Wollen Sie einen Artikel ansehen oder Ihre
Anfrage modifizieren?*
[presenting new list and asking the user how to proceed]
U: decides to go back and retrieve the previous set of results
S: *Moment. Hier ist die alte Ergebnisliste.*
[presenting old list]
U: chooses to look at the biography of "Peter Behrens", clicks on save and exit
S: *Sie wollen wirklich das System verlassen?"*
[asking if user really wants to quit]
U: confirms
S: *Unter welchen Namen wollen Sie die Artikel über Mackintosh und Behrens
speichern?*
[asking user to enter file names for the articles to be saved]
U: types in file names
S: *Vielen Dank und auf Wiedersehen!*

The remainder of this article is organized as follows. First, we discuss the
state-of-the-art in speech production (Section 2) and briefly describe the application context of human-machine dialogue in information seeking processes
(Section 3). Then we discuss the potential of taking a natural language generation perspective on speech production (Section 4). Focusing on the assignment
of intonation contours, Section 5 presents a functionally and contextually motivated model of speech production in dialogue that unites natural language
generation, dialogue modeling, and speech production. Section 6 concludes with
a summary of the arguments brought forward in the course of this paper and
points out a number of open issues we currently investigate.

2 Speech Production

In spoken information-seeking dialogues, intonation is often the only means to
distinguish between different dialogue acts, thus making the selection of the appropriate intonation crucial to the success of the dialogue (see e.g., Prevost &
Steedman, 1994 for English). To illustrate this point, imagine an information retrieval dialogue where the user wants to know about a particular painter. At some
point in the interaction, the system produces a sentence like *Sie wollen Infor-
mationen über die Werke von Picasso* ("You want information about the works
of Picasso"). Depending on the intonation, this utterance can be interpreted
as a statement, i.e., as giving information, or as a question, i.e., as requesting
information from the user, for example, in a clarification dialogue.

Current German speech synthesizers can support sophisticated variation of intonation, but no existing system provides the semantic nor pragmatic guidance necessary for selecting intonation appropriately. Traditional *text-to-speech* systems (e.g., Hemert, Adriaens-Porzig, & Adriaens, 1987; Huber *et al.*, 1987; Olaszy, Gordos, & Nemeth, 1992) perform a linguistic analysis of the written text and use the resulting syntactic structure to control prosodic features. *Concept-to-speech* systems (see e.g., Dorffner, Buchberger, & Kommenda, 1990) synthesize from a pre-linguistic conceptual structure but only use this structure to avoid the problem that syntactic analysis produces incomplete and multiple structures. *Discourse model based systems* (see e.g., Hirschberg, 1992; Prevost & Steedman, 1994) also start from a syntactic structure, and hence the discourse information is restricted to those kinds that can be reconstructed from that structure.

Moreover, the above mentioned systems assume a one-to-one mapping between syntactic structure and intonational features. Therefore, they cannot account for situations in which the same syntactic structure can be realized with differing intonation. Assuming that intonation is more than the mere reflection of the surface linguistic form (see Halliday, 1967; Pheby, 1969), and further, that intonation is selected to express particular communicative goals and intentions, we argue that an effective control of intonation requires synthesizing from *meanings* rather than word sequences. This is acknowledged in the SYN-PHONICS system (see Abb *et al.*, 1996), which allows prosodic features to be controlled by various factors other than syntax, e.g., by the information structure (focus–background or topic–comment structure). The function of intonation is, however, still restricted to grammatical function, more specifically *textual* function, without considering *interpersonal aspects*, i.e., communicative goals and speaker attitudes (see Halliday, 1967). Yet, in the context of generating speech in information-seeking dialogues, where intonational features are often the only means to signal a type of dialogue act, these aspects have to be taken into account.

Furthermore, in a dialogue situation it is not sufficient to look at isolated sentences; instead, one has to consider an utterance as part of a larger interaction. Intonation is not only used to mark sentence-internal information structure, but additionally, it is employed in the management of the communicative demands of interaction partners. Therefore, we also have to consider the function of intonation with respect to the context, taking into account the discourse (dialogue) history (see also Fawcett, van der Mije, & van Wissen, 1988). Intonation as realization of interactional features thus draws on dialogue and user models as additional sources of constraints.

3 Human-Machine Dialogue

In the SPEAK! project we employ a speech act oriented dialogue model that was originally developed in support of multimodal human-machine interaction. Its central part, the COR (COnversational Roles) model (Sitter & Stein, 1996), regards information-seeking dialogue as negotiation whose basic units are linguistic and non-linguistic communicative acts. COR is a general dynamic model of information-seeking dialogue and has been combined with a model of global dialogue structure represented by *scripts* for information-seeking strategies (Belkin *et al.*, 1995; Stein *et al.*, 1997). In SPEAK!, we use both scripts and a modified version of the COR model (Hagen & Stein, 1996).

COR covers all kinds of information-seeking dialogues, e.g., simple factual information dialogues as well as more complex negotiations. Focusing on *information* does not exclude *action*. In information retrieval systems, for example, the interactants do not only exchange information but may also negotiate for actions or goods and services, such as terminology support, functions for data manipulation, etc. Since moves for action and moves for information are closely coupled, the COR model does not distinguish between them (see Figure 2).

Comparable models of exchange have been developed, for example, by Bilange (1991), Fawcett et al. (1988), O'Donnell (1990) Traum & Hinkelman (1992). However, they address specific kinds of "information dialogue" (Bunt, 1989), i.e., exchange of factual information within well defined task-settings. Tasks and goals are usually not that well defined in information retrieval systems, which provide access to large textual information sources or multimedia databases, and the dialogue model must account must account for these characteristics[6].

COR uses recursive state-transition networks as representation formalism[7], i.e., a *dialogue* network and several *move* networks (for details see Hagen & Stein, 1996). Transitions in the dialogue net are moves, e.g., **Request, Offer, Reject, Withdraw, Accept, Inform, Evaluate** (see Figure 2). Moves are also represented as recursive transition networks consisting of several components. The prominent transition (the "nucleus") of each move is an atomic dialogue *act* which has the same illocutionary point as the superordinate move and is thus assigned the same speech act type, e.g., **Request**. The atomic act may be followed or preceded by optional "satellites" such as other *moves* (to supply additional information related to the atomic act), and sub-*dialogues* for clarification (to seek additional information).

Figure 2 contains the definition of types of some prototypical dialogue moves and the corresponding atomic acts. It also describes the differentiation between initiating vs. responding and expected vs. unexpected moves, which are a reflection of speaker–hearer role relations and expectations[8]. The possible sequences

---

[6] Dahlbäck (this volume) mentions the influence of different task-settings on the dialogue structure, also using the term "dialogue-task distance" in this context.

[7] The representation formalism for the linguistic resources (semantics, grammar) are *system networks*, the common representation formalism in Systemic Functional Linguistics (Halliday, 1985). Since the COR model was developed independently of the KOMET-PENMAN generator, representation not being uniform has simply historical reasons.

[8] There are a number of other unexpected dialogue contributions not displayed in Figure 2, such as moves and subdialogues which are embedded in other moves. An atomic request act of S, for instance, may be followed by a statement of S (e.g., an explanation or other background information), or K may initiate a clarification subdialogue (e.g., asking for additional information on S's request).

Dialogue Move			Follow-up Move		
Name	Transition	Definition	Expected	Unexpected	Transition
request (S,K)	1 → 2	S wants: K does A	promise (K,S)	reject-request (K,S)	2 → 1 or 7
				withdraw-request (S,K)	2 → 1 or 7
offer (K,S)	1 → 2'	K intends: K does A	accept (S,K)	reject-offer (S,K)	2' → 1 or 7'
				withdraw-offer (K,S)	2' → 1 or 7'
promise (K,S)	2 → 3	K intends: K does A*	inform (K,S)	withdraw-promise (K,S)	3 → 1 or 8
accept (S,K)	2' → 3	S wants: K does A*	inform (K,S)	withdraw-accept (S,K)	3 → 1 or 8
inform (K,S)	3 → 4	K believes: P	evaluate (S,K)	evaluate & quit (S,K)	4 → 5
evaluate (S,K)	4 → 1	S believes: P	request (S,K)	meta-dialogue (–,–)	1 → 1
			offer (K,S)	withdraw (–,–)	1 → 6
reject-request (K,S)	2 → 1	K intends: [not (K does A*)]	request (S,K)	meta-dialogue (–,–)	1 → 1
			offer (K,S)	withdraw (–,–)	1 → 6
reject-offer (S,K)	2' → 1	S wants: [not (K does A*)]	request (S,K)	meta-dialogue (–,–)	1 → 1
			offer (K,S)	withdraw (–,–)	1 → 6
reject-request (K,S)	2 → 7	K intends: [not (K does A*)]	— END —		
......	......	......	......	......	......

Notation: Order of parameters: first = speaker; second = addressee; S = information Seeker; K = information Knower; – = S or K; P= Proposition; A = Action (as defined in this move); A* = A adopted from preceeding move; numbers indicate dialogue states: states 1–4 build a dialogue cycle; states 5–8 are terminal states.

Fig. 2. COR dialogue moves and role assignments

of moves in a dialogue can be followed moving through the table across and down. Each dialogue cycle is either initiated by a Request of the information seeker or by an Offer of the information knower. These initiating moves and the responding move Inform have a task-oriented function in that they change the semantic context (cf. Bunt, 1996), whereas all of the other moves have dialogue control functions.

Applying the COR model for monitoring real dialogues, a *hierarchical dialogue history* (Hagen & Stein, 1996) can dynamically be constructed, where the actual utterances are represented as parts of a larger interaction: the atomic acts are sub-elements of moves, moves are embedded in other moves and/or sub-dialogues, etc. This knowledge about dialogue structures and related role expectations may be shared by the dialogue manager and other system components to generate coherent, context-dependent system reactions.

4 A Generation Perspective on Speech Production in Dialogue

In Section 2 we noted that in most speech production systems the root of the problem of making appropriate intonation assignments is insufficient knowledge about the sources of constraints on intonation. If meaning as constraining factor is considered at all, it is propositional meaning and possibly some aspects of textual meaning that are acknowledged as influencing intonational selection. While this is a step in the right direction, either it remains difficult to derive these meanings from a string (in the *text-to-speech* analysis view) or the kinds of meanings included remain functionally under-diversified (as in the *concept-to-speech* view). What is left unrepresented is the *interpersonal* aspect, such as speech function, speakers' attitudes, hearers' expectations, and speaker–hearer role relations in the discourse.

In a natural language generation perspective, a differentiated representation of linguistic meaning is usually an integral part. Also, in full-fledged generation systems, it is essential to have *contextual* knowledge about the *type of discourse* or *genre* that is to be generated[9] and about the *communicative goals* of a piece of discourse to be produced. Without this knowledge, grammatical and lexical expressions will be severely under-constrained and a generation grammar will fail to produce any sensible output. In the same way, in the context of speech production, intonation selection is under-constrained if it is not explicitly related to its *meanings in context*. If we consider intonation as a linguistic resource just like we do grammar and lexis, we can apply the generation scheme and explore the sources of constraints on intonation selection in the same way as we do with grammar and lexis.

In the present scenario, we suggest that our dialogue model can provide the extra-linguistic contextual information needed to satisfy the informational needs of the natural language/speech generation component. It does so in the following ways: First, the dialogue model provides information about *genre* in that it identifies move combinations typical of information-seeking dialogues (called *scripts*; see Section 3). The sequential ordering of moves is inherent to the model (see Figure 2 for an overview of move types, their definitions and possible follow-up moves). We call this *communicative context.* Second, the dialogue history built up during the traversal of the COR dialogue and move networks provides information about the *immediate context* in which an utterance occurs. We will see below (Section 5.2) that this is vital for intonation selection. Third, the categories the dialogue model employs are essentially *interpersonal* in nature. They have been

---

[9] See Hasan, 1978 for the coinage of the term *genre.*

motivated by traditional speech act theory (Searle, 1979), on the one hand, and by observations of typical interactions in information-seeking dialogues (Belkin et al., 1995), on the other hand. There is a specifiable mapping between these and the interpersonal semantic categories we adopt in our linguistic modeling of speech functions (see Section 5.2).

For the purpose of assigning appropriate intonation contours, an integration of our dialogue model and natural language generation system, therefore, suggests itself. In the following section we discuss in more detail how we employ the dialogue component of the SPEAK! system to constrain semantic and grammatical choice including the selection of intonation in the German version of a Systemic Functional Linguistic (SFL) based generation system, the KOMET-PENMAN text generator (PENMAN Project, 1989; Bateman & Teich, 1995).

5 Integration of NL Generation, Dialogue and Speech

5.1 Intonation as Part of Generation Resources

From a *Systemic Functional Linguistics* viewpoint intonation is just one means among others—such as syntax and lexis—to realize grammatical selection (see Halliday, 1985; Matthiessen, 1995). This implies that choices underlying the realization of intonation may be organized in exactly the same way as other choices in the grammar (see Fawcett, 1990; Halliday, 1967). Hence, the intonational control required for speech generation in a dialogue system has been built into the existing generation grammar for German (Grote, 1995; Teich, 1992). The added discriminations—i.e., more delicate choices in those areas of the grammar where intonational distinctions exist—are constraints on the specification of an appropriate intonation rather than constraints on syntactic structure.

There are three distinct kinds of phonological categories, i.e., *tone group, tonic syllable* and *tone* (see e.g., Bierwisch, 1973; Pheby, 1969), the determination of which requires a representation of:

- **Tonality:** The division of a text into a certain number of tone groups
- **Tonicity:** The placing of the tonic element within the tone group
- **Tone:** The choice of a tone for each tone group.

Choices in **tonality** and **tonicity** lead to an information constituent structure independent of the grammatical constituency, whereas choices in **tone** result in the assignment of a tone contour for each identified tone group in an utterance. The choices and hence the assignment of intonation features are constrained by knowledge of the intended textual and interpersonal meaning of the text.

While textual meaning is encoded by tonality and tonicity, interpersonal meaning is mostly encoded by tone. Following Pheby (1969), we assume five *primary tones*, plus a number of *secondary tones* that are necessary for the description of German intonation contours. These tones are: *fall* (tone1), *rise* (tone2), *progredient* (tone3), *fall-rise* (tone4), *rise-fall* (tone5), where the first four can be further differentiated into secondary *a* and *b* tones.[10]

The primary tones are the undifferentiated variants, whereas the secondary tones are interpreted as realizing additional meaning. The (primary) tone selection in a tone group serves to realize a number of distinctions in speech functions. For instance, depending on the tone contour selected, the system output *//sie wollen die biographie sehen//* ("You want to look at the biography") can be either interpreted as a question (tone **2a**) or a statement (tone **1a**). The (secondary) tone is conditioned by attitudinal options such as the speaker's attitude towards the proposition being expressed (surprise, reservation, ...), what answer is being expected, emphasis on the proposition, etc., referred to as *key*. Consider the following example taken from an information-seeking dialogue: The computer has retrieved an answer to a query, and this answer is presented visually to the user. As a default, the system would generate a neutral statement choosing tone **1a** to accompany the presentation, as in *//1a die ergebnisse sind unten DARgestellt//*[11] ("The results are presented below"). If, however, the results had so far been presented at a different position on the screen, the system would generate tone **1b** in order to place special emphasis on the statement: *// **1b** die ergebnisse sind UNTEN dargestellt//*.

The interpersonal meaning encoded by tone is speech function (statement, question, command, offer) in the first instance and speaker's attitude in the second instance. In the grammar, these are realized in *mood* and *key*. In SFL the *key* options are refinements of *mood* and have accordingly been implemented as an integral part of the grammar.

5.2 From Extra-Linguistic Context to Speech

With the intonational resources in place as described in the preceding section, we now proceed to our proposal of integrating generation, dialogue, and speech.

It is generally acknowledged that intonation is in the first instance the realization of interpersonal-semantic *speech function* (command, statement, question, offer). In our SFL based generation perspective, speech function meanings are accommodated on the semantic stratum of the generation resources and related to the grammar by a realization relation. Speech function is typically recoded grammatically in sentence mood (declarative, interrogative, imperative). However, the relation between speech function and mood is not one-to-one, and the

---

[10] The criteria for the distinction of primary tones is the *type* of the tone movement, for instance rising or falling tone contour, whereas the *degree* of the movement, i.e., whether it is strong or weak in expression, is considered to be a variation within a given tone contour.

[11] In this paper, the following notational conventions hold: // marks tone group boundaries, CAPITAL LETTERS are used to mark the tonic element of a tone group. Numbers following the // at the beginning of a tone group indicate the type of tone contour. The MULTIVOX text-to-speech system has been extended to interpret these notational conventions (Olaszy *et al.*, 1995).

speech function 'offer' has no prototypical grammaticalization at all. Imperative, interrogative, and declarative mood may encode a command, for example, *Speichern Sie den Artikel!* ("Save the article!"), *Würden Sie den Artikel speichern, bitte?* ("Would you please save the article?"), *Der Artikel sollte gespeichert werden.* ("The article should be saved."). This means that the mapping between speech function and mood must be constrained by some other, additional factors—and so must be the mapping to the *key* options opened up by mood selection which are realized in intonation (tone).

When we look at dialogue—as opposed to monologue—the other factors that come into play are the *type of dialogue move* to be verbalized and the *history of the dialogue* taking place. For instance, the genre of information-seeking, human-machine dialogue is characterized by certain genre-specific types of moves (see Section 3). A typical move in this genre is the **Request** move, which contains as its nucleus a simplex or atomic **Request** act. In terms of speech function, such a **Request** act is often a question. The **Request**–question correlation in the kind of dialogue we are dealing with here constrains the choice of mood to *interrogative* or *declarative*, e.g., (1) *Was möchten Sie sehen* ("What would you like to look at") (interrogative)— (2) *Sie wollen die Artikel über Picasso sehen* ("You want to look at the articles about Picasso") (declarative). So, in information-seeking dialogues, the type of move largely constrains the selection of speech function, but it only partially constrains the mapping of speech function and mood.

Deciding between declarative and interrogative as linguistic realization of a **Request** requires information about the *immediate context* of the utterance, i.e., about the dialogue history. It is in the dialogue history that speaker's attitudes and intentions and hearer's expectations are implicitly encoded. The area in the grammar recoding this kind of interpersonal information is *key*. The *key* systems are subsystems of the basic *mood* options (see Section 5.1) and realized by tone. Consider the contexts in which (1) or (2) would be appropriate: (1) would typically be used as an *initiating* act of an exchange, where there is no immediately preceding context—the speaker's attitude is essentially neutral and tone 1 is appropriate. (2) would typically be used in an exchange as the realization of a *responding* act. In terms of the dialogue model, (2) would be a possible realization of a **Request** initiating a clarification dialogue embedded in an **Inform** or **Request** move—the speaker wants to make sure she has understood correctly, and tone 2 is the appropriate intonation.

For the representation of constraints between *dialogue move* and *speech function* on the side of interpersonal semantics and *mood* and *key* on the part of the grammar, the *type* of dialogue move in context (the dialogue history) suggests itself as the ultimate constraint on tone selection. Given that all of the three parameters mentioned (dialogue move type/dialogue history, speech function, mood/key) are logically independent and that different combinations of them go together with different selections of tone, an organization of these parameters in terms of stratification appears appropriate, since it provides the required flexibility in the mapping of the different categories. Such an organization is, for instance, proposed in systemic functional work on interaction and dialogue (e.g., by Berry, 1981; Martin, 1992; Ventola, 1987 without considering intonation, and Fawcett, 1990 including intonation).

In the systemic functional model, the strata assumed are context (extra-linguistic), semantics and grammar (linguistic). On the semantic stratum, general knowledge about interactions is located, described in terms of the *negotiation* and speech function networks (cf. Martin, 1992). A pass through the negotiation network results in a syntagmatic structure of an interaction called *exchange structure*. An exchange structure consists of *moves* which are the units for which the *speech function* network holds. The *mood* and *key* systems represent the grammatical realization of a move (given that move is realized as a clause).

The ultimate constraint on the selection of features in this interpersonal semantics and the grammar is the information located at the stratum of extra-linguistic context. This is knowledge about the type of discourse or *genre*. In the present scenario, this type of contextual knowledge is provided by the dialogue model (COR combined with *scripts*), reflecting the genre of *information-seeking human-machine dialogue*. Since the stratum of context is extra-linguistic, locating the dialogue model—which has originally not been designed to be a model of *linguistic* dialogue, but of retrieval dialogue in general—here is a straightforward step. Figure 3 gives a graphical overview of the stratified architecture suggested.

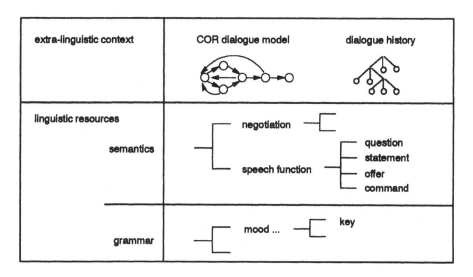

Fig. 3. A stratified model of resources constraining intonational choice

The general relation that holds between the strata is that of *realization*, which is implemented in KOMET-PENMAN-style generation as *inter-stratal preselection* (Bateman & Teich, 1995) and as *chooser-inquiry interface* (Matthiessen, 1988). The interstratal mappings can then be flexibly formulated depending on the dialogue history and the classification of dialogue moves (see again Figure 2), hence choice of intonation is sufficiently constrained. For a detailed account of the approach presenting the mood and key networks and providing a number of different sample dialogues and according intonational constraints see Teich *et al.* (1997).

6 Summary and Conclusions

In this paper we have argued that one of the shortcomings of state-of-the-art speech production systems is the selection of appropriate intonation contours. We have pointed out that the source of the problem is the analysis orientation of most speech production research. We suggested that it is worthwhile to more seriously attempt an integration of natural language and speech processing and consider the problem of intonation selection from the perspective of *natural language generation*. Taking such a perspective, we have presented a speech production approach that is embedded in a systemic-functionally based generation system, the KOMET-PENMAN system, used for the generation of spoken output in a human-machine dialogue scenario.

There is a number of open issues that we could not touch upon here which we are currently investigating. These include:

- further empirically backing up the intonational classification (Pheby, 1969) by acoustic analyses and auditive experiments, and, if necessary, revising it;
- carrying out more extensive analyses of human-machine dialogue using the SPEAK! system in order to validate the suggested model;
- adapting some of the text planning mechanisms we are currently working on for monologic text in the KOMET-PENMAN system for the mapping between the contextual and the semantic level;
- accounting also for the *textual* meanings of intonation, such as information structure and thematic progression, which are reflected in the selection of the tonic element (as for instance, suggested by Nakatani, 1995).

Dealing with these issues is a further necessary step to validate and complement what we call a *context-to-speech* approach, in which linguistic resources are sufficiently diversified to allow flexible intonation assignments and in which extra-linguistic information is taken into consideration as a major source of constraint on intonation selection.

References

[Abb *et al.*, 1996] Abb, B.; Günther, C.; Herweg, M.; Maienborn, C.; and Schopp, A. 1996. Incremental syntactic and phonological encoding – an outline of the SYN-PHONICS formulator. In Adorni, G., and Zock, M., eds., *Trends in Natural Language Generation — An Artificial Intelligence Perspective*. Berlin and New York: Springer-Verlag. 277–299.

[Bateman & Teich, 1995] Bateman, J. A., and Teich, E. 1995. Selective information presentation in an integrated publication system: An application of genre-driven text generation. *Information Processing & Management* 31(5):379–395.

[Belkin et al., 1995] Belkin, N. J.; Cool, C.; Stein, A.; and Thiel, U. 1995. Cases, scripts, and information seeking strategies: On the design of interactive information retrieval systems. *Expert Systems and Application* 9(3):379–395.

[Berry, 1981] Berry, M. 1981. Systemic linguistics and discourse analysis: A multi-layered approach to exchange structure. In Coulthard, M., and Montgomery, M., eds., *Studies in Discourse Analysis*. London: Routledge and Kegan Paul.

[Bierwisch, 1973] Bierwisch, M. 1973. Regeln für die Intonation deutscher Sätze. In *Studia Grammatica VII: Untersuchungen über Akzent und Intonation im Deutschen*. Berlin: Akademie Verlag. 99–201.

[Bilange, 1991] Bilange, E. 1991. A task independent oral dialogue model. In *Proceedings of the European Chapter of the ACL*, 83–87.

[Black & Campbell, 1995] Black, A., and Campbell, N. 1995. Predicting the intonation of discourse segments from examples in dialogue speech. In Dalsgaard, P.; Larsen, L. B.; Boves, L.; and Thomsen, I., eds., *Proceedings of the ESCA Workshop on Spoken Dialogue Systems—Theories and Applications (ETRW '95), Vigsø, Denmark*. Aalborg, Denmark: ESCA/Aalborg University. 197–200.

[Bunt, 1989] Bunt, H. C. 1989. Information dialogues as communicative action in relation to partner modeling and information processing. In Taylor, M. M.; Neel, F.; and Bouwhuis, D. G., eds., *The Structure of Multimodal Dialogue*. Amsterdam: North-Holland. 47–73.

[Bunt, 1996] Bunt, H. C. 1996. Interaction management functions and context representation requirements. In LuperFoy, S.; Nijholt, A.; and van Zanten, G., eds., *Dialogue Management in Natural Language Systems. Proceedings of the Eleventh Twente Workshop on Language Technology*, 187–198. Enschede, NL: Universiteit Twente.

[Callan, Croft, & Harding, 1992] Callan, J. P.; Croft, W. B.; and Harding, S. M. 1992. The INQUERY retrieval system. In *Proceedings of the 3rd International Conference on Database and Expert Systems Application*. Berlin and New York: Springer-Verlag. 78–83.

[Dahlbäck, 1997] Dahlbäck, N. 1997. Towards a dialogue taxonomy. In *this volume*.

[Dalsgaard et al., 1995] Dalsgaard, P.; Larsen, L. B.; Boves, L.; and Thomsen, I., eds. 1995. *Proceedings of the ESCA Workshop on Spoken Dialogue Systems—Theories and Applications (ETRW '95), Vigso, Denmark*. Aalborg, Denmark: ESCA/Aalborg University.

[Dorffner, Buchberger, & Kommenda, 1990] Dorffner, G.; Buchberger, E.; and Kommenda, M. 1990. Integrating stress and intonation into a concept-to-speech system. In *Proceedings of the 14th International Conference on Computational Linguistics (COLING '90)*, 89–94.

[Fawcett, van der Mije, & van Wissen, 1988] Fawcett, R. P.; van der Mije, A.; and van Wissen, C. 1988. Towards a systemic flowchart model for discourse. In *New Developments in Systemic Linguistics*. London: Pinter. 116–143.

[Fawcett, 1990] Fawcett, R. P. 1990. The computer generation of speech with discoursally and semantically motivated intonation. In *Proceedings of the 5th International Workshop on Natural Language Generation (INLG '90)*.

[Grote, 1995] Grote, B. 1995. Specifications of grammar/semantic extensions for inclusion of intonation within the KOMET grammar of German. COPERNICUS '93 Project No. 10393, SPEAK!, Deliverable R2.1.1.

[Hagen & Stein, 1996] Hagen, E., and Stein, A. 1996. Automatic generation of a complex dialogue history. In McCalla, G., ed., *Advances in Artificial Intelligence. Proceedings of the Eleventh Biennial of the Canadian Society for Computational Studies of Intelligence (AI '96)*. Berlin and New York: Springer-Verlag. 84–96.

[Halliday, 1967] Halliday, M. 1967. *Intonation and Grammar in British English*. The Hague: Mouton.

[Halliday, 1985] Halliday, M. 1985. *An Introduction to Functional Grammar*. London: Edward Arnold.

[Hasan, 1978] Hasan, R. 1978. Text in the systemic-functional model. In Dressler, W., ed., *Current Trends in Text Linguistics*. Berlin: de Gruyter. 228–246.

[Hemert, Adriaens-Porzig, & Adriaens, 1987] Hemert, J.; Adriaens-Porzig, U.; and Adriaens, L. 1987. Speech synthesis in the SPICOS project. In Tillmann, H., and Willee, G., eds., *Analyse und Synthese gesprochener Sprache. Jahrestagung der GLDV*. Hildesheim: Georg Olms. 34–39.

[Hirschberg, Nakatani, & Grosz, 1995] Hirschberg, J.; Nakatani, C.; and Grosz, B. 1995. Conveying discourse structure through intonation variation. In Dalsgaard, P.; Larsen, L.; Boves, L.; and Thomsen, I., eds., *Proceedings of the ESCA Workshop on Spoken Dialogue Systems—Theories and Applications (ETRW '95), Vigso, Denmark*. Aalborg, Denmark: ESCA/Aalborg University. 189–192.

[Hirschberg, 1992] Hirschberg, J. 1992. Using discourse context to guide pitch accent decisions in synthetic speech. In Bailly, G., and Benoit, C., eds., *Talking machines: Theory, Models and Design*. Amsterdam: North Holland. 367–376.

[Huber et al., 1987] Huber, K.; Hunker, H.; Pfister, B.; Russi, T.; and Traber, C. 1987. Sprachsynthese ab Text. In Tillmann, H. G., and Willee, G., eds., *Analyse und Synthese gesprochener Sprache. Jahrestagung der GLDV*. Hildesheim: Georg Olms. 26–33.

[LuperFoy, Nijholt, & van Zanten, 1996] LuperFoy, S.; Nijholt, A.; and van Zanten, G. V., eds. 1996. *Dialogue Management in Natural Language Systems. Proceedings of the Eleventh Twente Workshop on Language Technology*. Enschede, NL: Universiteit Twente.

[Martin, 1992] Martin, J. R. 1992. *English Text: System and Structure*. Amsterdam: Benjamins. chapter 7, 493–573.

[Matthiessen, 1988] Matthiessen, C. M. I. M. 1988. Semantics for a systemic grammar: The chooser and inquiry framework. In Benson, J.; Cummings, M.; and Greaves, W., eds., *Linguistics in a Systemic Perspective*. Amsterdam: Benjamins.

[Matthiessen, 1995] Matthiessen, C. M. I. M. 1995. *Lexicogrammatical Cartography: English Systems*. Tokyo: International Language Science Publishers.

[Nakatani, 1995] Nakatani, C. 1995. Discourse structural constraints on accent in narrative. In van Santen, J.; Sproat, R.; Olive, J.; and Hirschberg, J., eds., *Progress in Speech Synthesis*. Berlin and New York: Springer-Verlag.

[O'Donnell, 1990] O'Donnell, M. 1990. A dynamic model of exchange. *Word* 41(3):293–327.

[Olaszy et al., 1995] Olaszy, G.; Nemeth, G.; Tihanyi, A.; and Szentivanyi, G. 1995. Implementation of the interface language in the SPEAK! dialogue system. COPERNICUS '93 Project No. 10393, SPEAK!, Deliverable P2.3.1.

[Olaszy, Gordos, & Nemeth, 1992] Olaszy, G.; Gordos, G.; and Nemeth, G. 1992. The MULTIVOX multilingual text-to-speech converter. In Bailly, G., and Benoit, C., eds., *Talking Machines: Theory, Models and Design*. Amsterdam: North Holland. 385–411.

[PENMAN Project, 1989] PENMAN Project. 1989. PENMAN documentation: the Primer, the User Guide, the Reference Manual, and the Nigel manual. Technical report, University of Southern California/Information Sciences Institute, Marina del Rey, CA.

[Pheby, 1969] Pheby, J. 1969. *Intonation und Grammatik im Deutschen.* Berlin: Akademie-Verlag, (2nd. edition, 1980) edition.

[Prevost & Steedman, 1994] Prevost, S., and Steedman, M. 1994. Specifying intonation from context for speech synthesis. *Speech Communication* 15(1-2):139–153. Also available as http://xxx.lanl.gov/cmp-lg/9407015.

[Searle, 1979] Searle, J. R. 1979. *Expression and Meaning. Studies in the Theory of Speech Acts.* Cambridge, MA: Cambridge University Press. chapter A Taxonomy of Illocutionary Acts, 1–29.

[Sitter & Stein, 1996] Sitter, S., and Stein, A. 1996. Modeling information-seeking dialogues: The conversational roles (COR) model. *RIS: Review of Information Science* 1(1, Pilot Issue). Online Journal. Available from http://www.inf-wiss.uni-konstanz.de/RIS/.

[Stein et al., 1997] Stein, A.; Gulla, J. A.; Müller, A.; and Thiel, U. 1997. Conversational interaction for semantic access to multimedia information. In Maybury, M. T., ed., *Intelligent Multimedia Information Retrieval.* Menlo Park, CA: AAAI/The MIT Press. chapter 20. (in press).

[Teich et al., 1997] Teich, E.; Hagen, E.; Grote, B.; and Bateman, J. A. 1997. From communicative context to speech: Integrating dialogue processing, speech production and natural language generation. *Speech Communication.* (in press).

[Teich, 1992] Teich, E. 1992. KOMET: Grammar documentation. Technical Report, GMD-IPSI (Institut für integrierte Publikations- und Informationssysteme), Darmstadt.

[Traum & Hinkelman, 1992] Traum, D. R., and Hinkelman, E. 1992. Conversation acts in task-oriented spoken dialogue. *Computational Intelligence* 8(3):575–599.

[Ventola, 1987] Ventola, E. 1987. *The Structure of Social Interaction: A Systemic Approach to the Semiotics of Service Encounters.* London: Pinter.

Input Segmentation of Spontaneous Speech in JANUS: A Speech-to-Speech Translation System

Alon Lavie[1], Donna Gates[1], Noah Coccaro and Lori Levin[1]

Center for Machine Translation,
Carnegie Mellon University,
5000 Forbes Ave.,
Pittsburgh, PA 15213, USA

Abstract. JANUS is a multi-lingual speech-to-speech translation system designed to facilitate communication between two parties engaged in a spontaneous conversation in a limited domain. In this paper we describe how multi-level segmentation of single utterance turns improves translation quality and facilitates accurate translation in our system. We define the basic dialogue units that are handled by our system, and discuss the cues and methods employed by the system in segmenting the input utterance into such units. Utterance segmentation in our system is performed in a multi-level incremental fashion, partly prior and partly during analysis by the parser. The segmentation relies on a combination of acoustic, lexical, semantic and statistical knowledge sources, which are described in detail in the paper. We also discuss how our system is designed to disambiguate among alternative possible input segmentations.

1   Introduction

JANUS is a multi-lingual speech-to-speech translation system designed to facilitate communication between two parties engaged in a spontaneous conversation in a limited domain. It currently translates spoken conversations in which two people are scheduling a meeting with each other. The analysis of spontaneous speech requires dealing with problems such as speech disfluencies, looser notions of grammaticality and the lack of clearly marked sentence boundaries. These problems are further exacerbated by errors of the speech recognizer. In this paper we describe how multi-level segmentation of single utterance turns improves translation quality and facilitates accurate translation in our system. We define the basic dialogue units that are handled by our system, and discuss the cues and methods employed by the system in segmenting the input utterance into such units. Utterance segmentation in our system is performed in a multi-level incremental fashion, partly prior to and partly during analysis by the parser. The segmentation relies on a combination of acoustic, lexical, semantic and statistical knowledge sources, which are described in detail in the paper. We also discuss how our system is designed to disambiguate among alternative possible input segmentations.

The remainder of this paper is organized in the following way. We begin with an overview of the translation part of the JANUS system in Section 2.

In Section 3 we define the basic dialogue units which we model in our system, and describe how our system goes about translating such basic units. Section 4 discusses our initial input segmentation that is performed prior to parsing. Section 5 deals with parse-time segmentation of the input into basic dialogue units, and addresses the issue of disambiguation among alternative segmentations. In Section 6 we report our most recent results from an end-to-end translation evaluation with and without pre-segmented input. Finally, we present our summary and conclusions in Section 7.

2 System Overview

A diagram of the general architecture of the JANUS system is shown in Figure 1. The JANUS system is composed of three main components: a speech recognizer, a machine translation (MT) module and a speech synthesis module. The speech recognition component of the system is described elsewhere [10]. For speech synthesis, we use a commercially available speech synthesizer.

At the core of the system are two separate translation modules which operate independently. The first is the Generalized LR (GLR) module, designed to be more accurate. The second is the Phoenix module [5], designed to be more robust. Both modules follow an interlingua-based approach. In this paper, we focus on the GLR translation module. The results that will be reported in this paper will be based on the performance of the GLR module except where otherwise noted.

The source language input string is first analyzed by the GLR* parser [3][2]. Lexical analysis is provided by a morphological analyzer [4] based on Left Associative Morphology [1]. The parser uses a set of grammar rules in a unification-based formalism to produce a language-independent interlingua content representation in the form of a feature structure [8]. The parser is designed to be robust over spontaneous speech in that it skips parts of the utterance that it cannot incorporate into a well-formed interlingua. After parsing, the interlingua is augmented and completed by the discourse processor [6] where it is also assigned a speech-act, and then passed to a generation component [9], which produces an output string in the target language.

3 Semantic Dialogue Units for Speech Translation

JANUS is designed to translate spontaneously spoken dialogues between a pair of speakers. The current domain of coverage is appointment scheduling, where the two speakers have the task of scheduling a meeting. Each dialogue is a sequence of *turns* - the individual utterances exchanged between the speakers. Speech translation in the JANUS system is guided by the general principle that spoken utterances can be analyzed and translated as a sequential collection of semantic dialogue units (SDUs), each of which roughly corresponds to a speech act. SDUs are semantically coherent pieces of information that can be translated independently. The interlingua representation in our system was designed

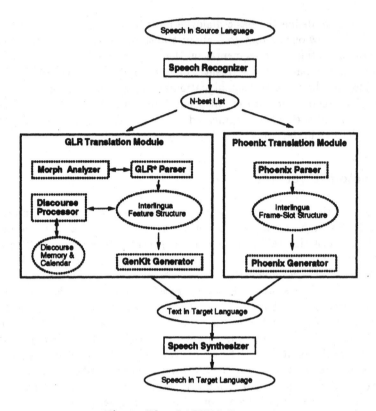

Fig. 1. The JANUS System

to capture meaning at the level of such SDUs. Each semantic dialogue unit is analyzed into an interlingua representation.

The analysis of a full utterance turn as a collection of semantic dialogue units requires the ability to correctly identify the boundaries between the units. This turns out to be a difficult yet crucial task, since translation accuracy greatly depends on correct segmentation. Utterance segmentation in our system is performed in a multi-level incremental fashion, partly prior to and partly during analysis by the parser. The segmentation relies on a combination of acoustic, lexical, semantic and statistical knowledge sources. These are described in detail in the following sections of the paper. The segmentation techniques are trained and developed on a corpus of transcribed dialogues with explicit markings of SDU boundaries.

3.1 Transcribing SDU boundaries

The system is trained and developed using a large corpus of recorded dialogues which are each transcribed. The recordings and their transcriptions are used in

```
mmxp_22_06: /h#/ si' [period] [seos] mira [period] [seos]
toda la man~ana estoy disponible [period] /h#/ [seos]
/eh/ y tambie'n los fin de semana [period] [seos]
si podri'a ser mejor un di'a de fin de semana [comma]
porque justo el once no puedo [period] [seos]
me es imposible [period] [seos] /gl/ [begin_simultaneous]
vos [end_simultaneous] pode's el fin de semana [quest] [seos]
```

Fig. 2. Transcription of a Spanish Utterance

the development and testing of various components of the system. Although the speech recordings have no explicit indications of SDU boundaries, we attempt to accurately detect and mark these boundaries in the transcriptions. While listening to the dialogues, the transcribers use acoustic signals as well as their own judgements of where sentential or fragment boundaries occur. Figure 2 shows an example of a transcribed utterance. The SDU boundaries are indicated with the transcription convention marking: *seos* (i.e., semantic end of segment).

3.2 Parsing SDUs

Our analysis grammars are designed to produce analyses for the variety of naturally occurring SDUs in spontaneously spoken dialogues in the scheduling domain. More often than not, SDUs do not correspond to grammatically complete sentences. SDUs often correspond to input fragments, clauses or phrases, which convey a semantically coherent piece of information. In particular, fragments such as time expressions often appear in isolation, and are allowed to form complete analyses and "float" to the top level of our grammars. The grammars are thus designed to analyze a full utterance turn as a sequence of analyzed SDUs. Figure 3 contains an example utterance in Spanish which demonstrates how a full utterance can consist of multiple SDUs.

4 Pre-parsing Segmentation

Segmentation decisions in our system can be most reliably performed during parsing, at which point multiple sources of knowledge can be applied. Nevertheless, we have discovered that there are several advantages to be gained from performing some amount of segmentation at the pre-parsing stage. The goal here is to detect highly confident SDU boundaries. We then pre-break the utterance at these points into sub-utterances that may still contain multiple SDUs. Each of these sub-utterances is then parsed separately.

Pre-parsing segmentation at SDU boundaries is determined using acoustic, statistical and lexical information. Segmentation at the pre-parsing stage has two

```
Transcription of Segmented Utterance Broken Down into SDUs:

(si1) (mira) (toda la man5ana estoy disponible)
(y tambie1n los fin de semana)
(si podri1a ser  mejor un di1a de fin
 de semana porque justo el once no puedo)
(me es imposible)
(vos pode1s el fin de semana)
```

```
Handmade Translation of the Utterance:

Yes. Look, all morning I'm free.
And also the weekend.  If it would be
better, a day on the weekend, because on
the eleventh I can't (meet). It is impossible
for me.  Can you (meet) on the weekend?
```

Fig. 3. Semantic Dialogue Units of the Utterance in Figure 2

main advantages. The first advantage is a potential increase in the parsability of the utterance. Although the GLR* parser is designed to skip over parts of the utterance that it cannot incorporate into a meaningful structure, only limited amounts of skipping are considered by the parser due to feasibility constraints. Often, a long utterance contains a long internal segment that is unparsable. If the parser does not manage to skip over the entire unparsable segment, a partial parse may be produced that covers either the portion of the utterance that preceded the unparsable segment, or the portion of the utterance that followed it, but not both. If, however, the utterance is pre-broken into several sub-utterances that are then parsed separately, there is a greater chance that parses will be produced for all portions of the original utterance.

The other potential advantage of pre-breaking an utterance is a significant reduction in ambiguity and subsequently a significant increase in efficiency. Considering all possible segmentations of the utterance into SDUs adds an enormous level of ambiguity to the parsing process. A long utterance may have hundreds of different ways in which it can be segmented into analyzable SDUs. This amount of ambiguity can be drastically reduced by determining some highly confident SDU boundaries in advance. Each of the sub-utterances passed on to the parser is then much smaller and has far fewer possible segmentations into SDUs that must be considered by the parser. Without pre-breaking, the unsegmented utterance in Figure 4 is parsable, but requires 113 seconds. Pre-breaking produces the set of three sub-utterances shown in example (2) of Figure 4. Parsing the three sub-utterances in sequence requires only 32 seconds (less than a third of the time) and yields the parser segmentation (3) and translations (4) shown at the end of Figure 4.

(1) **Unsegmented Speech Recognition:**

    (%noise% si1 mira toda la man5ana estoy disponible %noise%
    %noise% y tambie1n el fin de semana si podri1a hacer mejor
    un di1a fin de semana porque justo el once no puedo me es
    imposible va a poder fin de semana %noise%)

(2) **Pre-broken Speech Recognition:**

    (si1)
    (mira toda la man5ana estoy disponible
    %noise% %noise% y tambie1n el fin de semana)
    (si podri1a hacer mejor un di1a fin de semana)
    (porque justo el once no puedo me es imposible
    va a poder fin de semana)

(3) **Parser SDU Segmentation (of Pre-broken Input):**

    (((si1))
    ((mira) (toda la man5ana estoy disponible) (y tambie1n)
    (el fin de semana))
    ((si podri1a hacer mejor un di1a fin de semana))
    ((porque el once no puedo) (me es imposible)
    (va a poder fin de semana)))

(4) **Actual Translation:**

    "yes --- Look all morning is good for me -- and also --
    the weekend --- If a day weekend is better --- because
    on the eleventh I can't meet -- That is bad for me --
    can meet on weekend"

Fig. 4. Efficiency Effect of Pre-breaking on a Spanish Utterance

4.1 Acoustic Cues for SDU Boundaries

The first source of information for our pre-breaking procedure is acoustic information supplied by the speech recognizer. We found that some acoustic cues have a very high probability of occurring at SDU boundaries. To a certain extent, these cues are language-dependent. In general, long silences are a good indicator of an SDU boundary. After testing various combinations of noises within Spanish dialogues, we find that the following acoustic signals yield the best results for picking SDU boundaries: silences, two or more human noises in a row and three or more human or non-human noises in a row. It has been suggested in other work that pause units are good indicators of segment boundaries [7]. However, since multiple noises and silences inside an utterance are rare in our Spanish data, acoustic cues detect only a small fraction of the SDU boundaries. In addition, in at least one set of Spanish test dialogues recorded using six different speakers, we found no pause units at all. Thus, these acoustic cues alone are insufficient for solving the segmentation problem in Spanish.

4.2 Statistical Detection of SDU Boundaries

The second source of information for our pre-breaking procedure is a statistically trained confidence measure that attempts to capture the likelihood of an SDU boundary between any pair of words in an utterance. The likelihood of a boundary at a particular point in the utterance is estimated based on a window of four words surrounding the potential boundary location — the two words prior to the point in question and the two words following it. We denote a window of the words w_1, w_2, w_3, w_4 by $[w_1 w_2 \bullet w_3 w_4]$, where the potential SDU boundary being considered is between w_2 and w_3. There are three bigram frequencies that are relevant to the decision of whether or not an SDU boundary is likely at this point. These are:

1. $F([w_1 w_2 \bullet])$: the frequency of an SDU boundary being to the right of the bigram $[w_1 w_2]$.
2. $F([w_2 \bullet w_3])$: the frequency of an SDU boundary being between the bigram $[w_2 w_3]$.
3. $F([\bullet w_3 w_4])$: the frequency of an SDU boundary being to the left of the bigram $[w_3 w_4]$.

The bigram frequencies are estimated from a transcribed training set in which the SDU boundaries are explicitly marked. The frequencies are calculated from the number of times an SDU boundary appeared in the training data in conjunction with the appropriate bigrams. In other words, if $C([w_i w_j \bullet])$ is the number of times that a clause boundary appears to the right of the bigram $[w_i w_j]$ and $C([w_i w_j])$ is the total number of times that the bigram $[w_i w_j]$ appears in the training set, then

$$F([w_i w_j \bullet]) = \frac{C([w_i w_j \bullet])}{C([w_i w_j])}$$

$F([w_i \bullet w_j])$ and $F([\bullet w_i w_j])$ can be calculated in a similar fashion. However, for a given quadruple $[w_1 w_2 \bullet w_3 w_4]$, in order to determine whether the point in question is a reasonable place for breaking the utterance, we compute the following estimated frequency $\tilde{F}([w_1 w_2 \bullet w_3 w_4])$:

$$\tilde{F}([w_1 w_2 \bullet w_3 w_4]) = \frac{C([w_1 w_2 \bullet]) + C([w_2 \bullet w_3]) + C([\bullet w_3 w_4])}{C([w_1 w_2]) + C([w_2 w_3]) + C([w_3 w_4])}$$

This was shown to be more effective than the linear combination of the frequencies $F([w_1 w_2 \bullet])$, $F([w_2 \bullet w_3])$ and $F([\bullet w_3 w_4])$. The method we use is more effective because a bigram with a low frequency of appearance, for which we may not have sufficiently reliable information, is not counted as highly as the other factors.

When the calculated SDU boundary probability $\tilde{F}([w_1 w_2 \bullet w_3 w_4])$ exceeds a pre-determined threshold, the utterance will be segmented at this point. Setting the threshold for segmentation too low will result in high levels of segmentation, some of which are likely to be incorrect. Setting the threshold too high will result in ineffective segmentation. As already mentioned, even though pre-parsing segmentation can improve system efficiency and accuracy, segmentation decisions in our system can be done much more reliably at the parser level where syntactic and semantic rules help in determining valid SDU boundaries and prevent boundaries at syntactically or semantically ungrammatical locations in an utterance. Furthermore, an incorrect pre-parsing segmentation cannot be corrected at a later stage. For these reasons, we set the threshold for pre-parsing segmentation to a cautiously high value, so as to prevent incorrect segmentations as much as possible. The actual threshold was determined based on experimentation with several values over a large development set of dialogues. We determined the lowest possible threshold value that still did not produce bad incorrect segmentations. The statistical segmentation predictions were compared against the SDU boundary markers in the transcribed versions of the utterances to determine if a prediction was correct or false. The best threshold between 0 and 1.0 for pre-breaking was determined to be 0.6.

4.3 Lexical Cues for SDU Boundaries

The third source of information for our pre-breaking procedure is a set of lexical cues for SDU boundaries. These cues are language- and most likely domain-specific words or phrases that have been determined through linguistic analysis to have a very high likelihood of preceding or following an SDU boundary. While examining the results of the statistical pre-breaking, we noticed that there were phrases that almost always occurred at SDU boundaries. These phrases usually had high SDU boundary probabilities, but in some cases not high enough to exceed the threshold for SDU boundary prediction. We examined roughly 100 dialogues from the Scheduling domain to find phrases that commonly occur at SDU boundaries. We determined that the phrases *qué tal* ("how about..." or "how are you"), *qué te parece* ("how does ... seem to you"), and *si* ("if...") usually occur after an SDU boundary while *sí* ("yes") and *claro* ("clear") occur before an SDU boundary.

We modified the procedure for pre-breaking so that it could take into account these phrases. A small knowledge base of the phrases was created. The phrases alone do not trigger SDU boundary breaks. They are combined with the statistical component mentioned in the previous subsection. This is done by assigning the phrases a probability "boost" value. When the phrases are encountered by the pre-breaking program, the corresponding SDU boundary probability is incremented by the boost value of the phrase. After the optimal pre-breaking statistical threshold was determined to be 0.6, we experimented with several probability boost values for the phrases. These numbers were determined in the same manner as the best pre-breaking threshold. For phrases that occur after a boundary, we determined that the best probability boost value is

0.15. For phrases that occur before a boundary, the best probability boost value is 0.25. When one of the lexical phrases appears in an utterance, the increase in probability due to the boost value will usually increment the SDU boundary probability enough to exceed the threshold of 0.6. In cases where the probability still does not exceed the threshold, the original boundary probability is so low that it would be dangerous to break the utterance at that point. In such cases, we prefer to allow the parser to determine whether or not the point in question is in fact an SDU boundary.

4.4 Acoustic Cues vs. Statistical and Lexical Cues

As stated earlier, acoustic cues are not sufficient predictors of SDU boundaries in Spanish. We tested the translation performance of the GLR system on the output of the speech-recognizer by using just acoustic cues to determine the pre-parsing segmentation and then by using a combination of acoustic, statistical and lexical cues. When only using the acoustic cues, the in-domain translations of the output from the speech-recognizer had an acceptability rate of 51%. When combining acoustic, statistical and lexical cues, the translations of the same output from the speech-recognizer had an acceptability rate of 60%.

4.5 Performance Evaluation of Pre-parsing Segmentation

After combining the three knowledge sources of the pre-parsing segmentation procedure and empirically setting its parameters, we tested the performance of the procedure on an unseen test set. As explained earlier, the pre-parsing segmentor is designed to detect only highly confident SDU boundaries. However, it is crucial that it avoid incorrect segmentations. The parameters of the segmentor were tuned with these goals in mind.

The test set consisted of 103 utterances. The transcribed version of the test set indicated that the utterances contained 227 internal SDU boundaries that were candidates for detection. We evaluated the pre-parsing segmentation procedure on a transcribed version of the input, from which the SDU boundary indications were omitted, as well as on the actual speech recognizer output. On the transcribed input, the segmentation procedure detected 98 of the SDU boundaries (43%). On the speech recognized input, 129 SDU boundaries (57%) were detected. Some SDU boundaries suggested by the segmentation procedure were incorrect. 46 such incorrect boundaries were placed in the transcribed input. However, in only 3 cases did the incorrect segmentation adversely effect the translation produced by the system. Similarly, 60 incorrect segmentations were inserted in the speech recognized input, 19 of which had an adverse effect on translation. Most of the spurious segmentations occur around noise words and are inconsequential. Others occurred in segments with severe recognition errors.

5 Parse-time Segmentation and Disambiguation

Once the input utterance has been broken into chunks by our pre-parsing segmentation procedure, it is sent to the parser for analysis. Each utterance chunk corresponds to one or more SDUs. The GLR* parser analyzes each chunk separately, and must find the best way to segment each chunk into individual SDUs. Chunks that contain multiple SDUs can often be segmented in several different ways. As mentioned in the previous section, the number of possible SDU segmentations of a chunk greatly increases as a function of its length. In the example from Figure 4 one of the chunks that results from pre-parsing segmentation is *(porque justo el once no puedo me es imposible va a poder fin de semana)*. Because the grammar allows for sentence fragments, this chunk can be parsed into very small pieces such as *((porque) (justo) (el once) (no) (puedo) (me es) (imposible) (va a poder) (fin) de (semana))* or can be parsed into larger pieces such as *((porque justo el once no puedo) (me es imposible) (va a poder fin de semana))*. Many combinations of the smaller and larger pieces can also be parsed. This presents the parser with a significant additional level of ambiguity.

Even single SDUs may often have multiple analyses according to the grammar, and may thus prove to be ambiguous. *(Viernes dos)* ("Friday the second") may have additional parses such as ("Friday" "the second"), ("Friday at two") or ("Friday" "two o'clock"). Here the number *dos* can be a date or a time. The level of ambiguity introduced by chunks of multiple SDUs can drastically compound this problem. Dealing with such high levels of ambiguity is problematic from two different perspectives. The first is parser efficiency, which is directly correlated to the number of different analyses that must be considered in the course of parsing an input. The second perspective is the accuracy of the selected parse result. The greater the amount of ambiguity, the more difficult it is for the parser to apply its disambiguation methods successfully, so that the most "correct" analysis is chosen. The task of finding the "best" segmentation is therefore an integral part of the larger parse disambiguation process.

During parsing, the early pruning of ambiguities that correspond to chunk segmentations that are unlikely to be correct can result in a dramatic reduction in the level of ambiguity facing the parser. This can result in a significant improvement in both parser efficiency and accuracy.

5.1 The Fragmentation Counter Feature

Because our grammar is designed to be able to analyze fragments as first class SDUs, it is often the case that an input chunk can be analyzed both as a single SDU as well as a sequence of smaller fragment SDUs. In most cases, when such a choice exists the least fragmented analysis corresponds to the most semantically coherent representation. We therefore developed a mechanism for representing the amount of fragmentation in an analysis, so that less fragmented analyses could be easily identified.

The fragmentation of an analysis is reflected via a special "counter" slot in the output of the parser. The value of the counter slot is determined by explicit settings in the grammar rules. This is done by unification equations in the grammar rules that set the value of the counter slot in the feature structure corresponding the the left-hand side of the rule. In this way, the counter slot can either be set to some desired value, or assigned a value that is a function of counter slot values of constituents on the right-hand side of the rule.

By assigning counter slot values to the feature structures produced by rules of the grammar, the grammar writer can explicitly express the expected measure of fragmentation that is associated with a particular grammar rule. For example, rules that combine fragments in less structured ways can be associated with higher counter values. As a result, analyses that are constructed using such rules will have higher counter values than those constructed with more structurally "grammatical" rules, reflecting the fact that they are more fragmented. In particular, the high level grammar rules that chain together SDU-level analyses can sum the fragmentation counter values of the individual SDU analyses that are being chained together.

5.2 Pruning Analyses Using Fragmentation Counters

The preference for a less fragmented analysis is realized by comparing the different analyses of SDU chains as they are being constructed, and pruning out all analyses that are not minimal in their fragmentation values. The pruning heuristic is implemented as a procedure that is invoked along with the grammar rule that combines a new SDU analysis with a list of prior analyzed SDUs. The feature structure associated with the list of prior analyzed SDUs is pruned in a way that preserves only values that correspond to the minimum fragmentation. The feature structure of the new SDU is then combined only with these selected values.

Since the SDU combining grammar rule is invoked at each point where a part of the input utterance may be analyzed as a separate SDU, the pruning procedure incrementally restricts the parses being considered throughout the parsing of the input utterance. This results in a substantial decrease in the total number of ambiguous analyses produced by the parser for the given utterance, as well as a significant reduction in the amount of time and space used by the parser in the course of parsing the utterance.

5.3 Pruning Analyses Using Statistical Information

In addition to the fragmentation pruning, we use a statistical method that aims to prevent the parser from considering SDU boundaries at points in the utterance in which they are unlikely to appear. This is done using the same statistical information about the SDU boundary likelihood that is used for utterance pre-breaking. However, whereas in the pre-breaking process we attempt to detect locations in the utterance where an SDU boundary is likely to occur, within the parser we are attempting to predict the opposite, i.e., locations in which SDU boundaries are *unlikely*.

The likelihood of an SDU boundary is computed in the same fashion as previously described in Section 4. However, the information is now used differently. The procedure that calculates the SDU boundary likelihood is called by a special rule within the grammar, which is invoked whenever the parser completes a partial analysis that may correspond to a complete SDU. This grammar rule is allowed to succeed only if the point where the sentence ends is a statistically reasonable point to break the utterance. Should the rule fail, the parser will be prevented from pursuing a parse in which the following words in the utterance are interpreted as a new SDU. In order for the grammar rule to succeed, the computed boundary probability must be greater than a threshold set in advance. The value of the threshold was set empirically so as to try and obtain as much pruning as possible, while not pruning out correct SDU segmentations. It is currently set to 0.03 for both Spanish and English.

To test the effectiveness of the statistical method of pruning out analyses, we compared the results of parsing an English test set of 100 utterances, both with and without statistical pruning. Using the statistical pruning resulted in an overall decrease of about 30% in parsing time. A comparison of the parse analyses selected by the parser showed that with statistical pruning, the parser selected a better parse for 15 utterances, while for 7 utterances a worse parse was selected. Although the seven bad cases are a result of missed SDU boundaries, the 15 good cases are a result of the parser selecting a better SDU segmentation, due to the fact that analyses with incorrect SDU boundaries were statistically pruned out.

5.4   Parse Disambiguation

All SDU segmentations allowed by the grammar that are not pruned out by the methods previously described are represented in the collection of analyses that are output by the parser. A parse disambiguation procedure is then responsible for the task of selecting the "best" analysis from among this set. Implicitly this includes selecting the "best" SDU segmentation of the utterance.

Disambiguation in GLR* is done using a collection of parse evaluation measures which are combined into an integrated heuristic for evaluating and ranking the parses produced by the parser. Each evaluation measure is a penalty function, which assigns a penalty score to each of the alternative analyses, according to its desirability. The penalty scores are then combined into a single score using a linear combination.

The parser currently combines three penalty scores. The first is a skip penalty that is a function of the words of the utterance that were not parsed in the course of creating the particular analysis. Different analyses may correspond to different skipped portions of the utterance. The penalty for skipping a word is a function of the word's saliency in the scheduling domain. Highly salient words receive a high skip penalty. Analyses that skip fewer words, or words with lower saliencies are preferable, and thus receive lower penalties.

The fragmentation counter attached to the analysis is used as a second penalty score. As mentioned earlier, the value of the fragmentation counter slot

In Domain (248 SDUs)

	without pre-breaking	with pre-breaking
GLR	36%	54%
Phoenix	49%	52%

Fig. 5. Results of a translation evaluation with and without pre-broken speech-recognition output

reflects the amount of fragmentation of the analysis. For each of the parsable subsets of the utterance considered by the parser, pruning using fragmentation counters results in analyses that are minimal in the number of SDUs. In the disambiguation stage, where analyses of different parsable subsets are compared, the fragmentation counter is used as a penalty score, so as to once again reflect the preference for analyses that correspond to fewer SDUs.

The third penalty score is based on a statistical disambiguation module that is attached to the parser. The statistical framework is one in which shift and reduce actions of the LR parsing tables are directly augmented with probabilities. Training of the probabilities is performed on a set of disambiguated parses. The probabilities of the parse actions induce statistical scores on alternative parse trees, which are then used for disambiguation. Statistical disambiguation can capture structural preferences in the training data. This will usually create a bias toward structures that correspond to SDU segmentations that are more likely to be correct.

6 Results

In order to test the effect of using pre-breaking on the output of the speech recognizer, we performed an end-to-end translation evaluation on a set of three unseen Spanish dialogues consisting of 103 utterances. These dialogues were never used for system development or training. The dialogues were first translated by both modules from unsegmented speech output and then from automatically segmented speech output. The results reported here are based on the percentage of acceptable translations of the 248 in-domain SDUs from the test set. These translations were scored by an objective independent scorer. As can be seen in Figure 5, the accuracy of the GLR translation module increased significantly, from 36% to 54%. The Phoenix module is much less sensitive to the effects of pre-parsing segmentation. Thus, on unbroken utterances, Phoenix significantly out-performs GLR. However, pre-parsing segmentation results in a minor improvement in translation accuracy for the Phoenix translation module as well.

7 Summary and Conclusions

Accurate speech translation in JANUS requires that an input utterance be correctly segmented into semantic dialogue units. We achieve this task using a combination of acoustic, statistical, lexical and semantic information, which is applied in two stages, prior to parsing and during parsing. Pre-parsing segmentation is advantageous because it increases the robustness of the parser to unparsable segments in the input utterance and significantly reduces the amount of segmentation ambiguity presented to the parser. However, accurate segmentation is performed during parse-time, when semantic and grammatical constraints can be applied. Pruning heuristics allow the parser to ignore segmentations that are unlikely to be correct. This restricts the set of possible analyses passed on for disambiguation. The disambiguation process subsequently selects the analysis deemed most correct.

References

1. R. Hausser. Principles of Computational Morphology. Technical Report, Laboratory for Computational Linguistics, Carnegie Mellon University, Pittsburgh, PA, 1989.
2. A. Lavie. An Integrated Heuristic Scheme for Partial Parse Evaluation, Proceedings of the 32nd Annual Meeting of the ACL (ACL-94), Las Cruces, New Mexico, June 1994.
3. A. Lavie and M. Tomita. GLR* - An Efficient Noise Skipping Parsing Algorithm for Context Free Grammars, *Proceedings of the third International Workshop on Parsing Technologies (IWPT-93), Tilburg, The Netherlands, August 1993.*
4. L. Levin, D. Evans, and D. Gates. The ALICE System: A Workbench for Learning and Using Language. *Computer Assisted Language Instruction Consortium (CALICO) Journal*, Autumn 1991, 27–56.
5. L. Mayfield, M. Gavaldà, Y-H. Seo, B. Suhm, W. Ward, A. Waibel. Parsing Real Input in JANUS: a Concept-Based Approach. In *Proceedings of TMI 95.*
6. C. P. Rosé, B. Di Eugenio, L. S. Levin, and C. Van Ess-Dykema. Discourse processing of dialogues with multiple threads. In *Proceedings of ACL'95, Boston, MA*, 1995.
7. M. Seligman, J. Hosaka, and H. Singer: "Pause Units" and Analysis of Spontaneous Japanese Dialogues: Preliminary Studies This volume, 1997.
8. S. M. Shieber. *An Introduction to Unification-Based Approaches to Grammar*, CSLI Lecture Notes, No. 4, 1986.
9. M. Tomita and E. H. Nyberg 3rd. Generation Kit and Transformation Kit, Version 3.2: User's Manual. Technical Report CMU-CMT-88-MEMO, Carnegie Mellon University, Pittsburgh, PA, October 1988.
10. M. Woszczyna, N. Aoki-Waibel, F. D. Buo, N. Coccaro, K. Horiguchi, T. Kemp, A. Lavie, A. McNair, T. Polzin, I. Rogina, C. P. Rosé, T. Schultz, B. Suhm, M. Tomita, and A. Waibel. JANUS-93: Towards Spontaneous Speech Translation. In *Proceedings of IEEE International Conference on Acoustics, Speech and Signal Processing (ICASSP'94)*, 1994.

"Pause Units" and Analysis of Spontaneous Japanese Dialogues: Preliminary Studies

Mark Seligman(1) Junko Hosaka(2) Harald Singer(3)

(1) 1100 West View Drive, Berkeley, CA 94705, USA: seligman@cerf.net

(2) University of Tuebingen, Tuebingen, Germany: hosaka@sfs.nphil.uni-tuebingen.de

(3) ATR Interpreting Telecommunications Research Labs, Hikaridai 2-2, Seika-cho, Soraku-gun, Kyoto 619-02, Japan: singer@itl.atr.co.jp

Abstract

We consider the use of natural pauses to aid analysis of spontaneous speech, studying four Japanese dialogues concerning a simulated direction-finding task. Using new techniques, we added to existing transcripts information concerning the placement and length of significant pauses within turns (breathing intervals of any length or silences longer than approximately 400 milliseconds). We then addressed four questions: (1) Are "pause units" (segments bounded by natural pauses) reliably shorter than utterances? The answer was Yes: on average, pause units in our corpus were on average 5.89 Japanese morphemes long, 60% the length of whole utterances, with much less variation. (2) Would hesitation expressions yield shorter units if used as alternate or additional boundaries? The answer was Not much, apparently because pauses and hesitation expressions often coincide. We found no combination of expressions which gave segments as much as one morpheme shorter than pause units on average. (3) How well-formed are pause units from a syntactic viewpoint? We manually judged that 90% of the pause units in our corpus could be parsed with standard Japanese grammars once hesitation expressions had been filtered from them. (4) Does translation by pause unit deserve further study? The answer was Yes, in that a majority of the pause units in four dialogues gave understandable translations into English when translated by hand. We are thus encouraged to further study a "divide and conquer" analysis strategy, in which parsing and perhaps translation of pause units is carried out before, or even without, attempts to create coherent analyses of entire utterances.

Introduction

It is widely believed that prosody can prove crucial for speech recognition and analysis of spontaneous speech. So far, however, effective demonstrations have been few. There are several aspects of prosody which might be exploited: pitch contours, rhythm, volume modulation, etc. Our suggestion here will be to focus on natural pauses as an aspect of prosody which is both important and relatively easy to detect automatically.

From a syntactic point of view, spontaneous speech is notoriously messy. Given the frequency of utterances which are not fully well-formed — which contain repairs, hesitations, and fragments — strategies for dividing and conquering utterances would be quite useful. We will suggest that natural pauses can play a part in such a strategy: that *pause units*, or segments within utterances bounded by natural pauses, can provide chunks which (1) are reliably shorter and less variable in length than entire utterances and (2) are relatively well-behaved internally from the syntactic viewpoint, though analysis of the relationships among them appears more problematic.

We address four questions specifically: (1) Are pause units reliably shorter than whole utterances? If they were not, they could hardly be useful in simplifying analysis. We find however, that in our corpus pause units are in fact about 60% the length of entire utterances, on the average, when measured in Japanese morphemes. The average length of pause units was 5.89 morphemes, as compared with 9.39 for whole utterances. Further, pause units are less variable in length than entire utterances: the standard deviation is 5.79 as compared with 12.97. (2) Would hesitations give even shorter, and thus perhaps even more manageable, segments if used as alternate or additional boundaries? The answer seems to be that because hesitations so often coincide with pause boundaries, the segments they mark out are nearly the same as the segments marked by pauses alone. We found no combination of expressions which gave segments as much as one morpheme shorter than pause units on average. (3) Is the syntax within pause units relatively manageable? A manual survey shows that some 90% of our pause units can be parsed using standard Japanese grammars; a variety of special problems, discussed below, appear in the remaining 10%. (4) Is translation of isolated pause units a possibility? This question was of special interest because our research was begun during the development of a speech translation system [Morimoto et al. 1993]. We found that a majority of the pause units in four dialogues gave understandable translations into English when translated by hand.

Our investigation began with transcriptions of four spontaneous Japanese dialogues concerning a simulated direction-finding task. The dialogues were carried out in the EMMI-ATR Environment for Multi-modal Interaction [Loken-Kim, Yato, Kurihara, Fais, and Furukawa 1993], [Furakawa, Yato, and Loken-Kim 1993], two using telephone connections only, and two employing onscreen graphics and video as well. In each 3-7 minute dialogue, a caller pretending to be at Kyoto station received from a pre-trained "agent" directions to a conference center and/or hotel. In the multimedia setup, both the caller and agent could draw on onscreen maps and exchange typed information. The four conversations were actually carried out by only two subjects — call them Caller1 and Caller2 — to enable comparison between each individual's performance in the telephone-only and multi-media setups. Morphologically tagged transcripts of the conversations were divided into turns by the transcriber, and included hesitation expressions and other natural speech features. Using techniques described in Section 1, we added information concerning the placement and length of significant pauses within turns, where "significant" pauses are breathing intervals of any length or silences longer than approximately 400 milliseconds.

We then studied the length (in Japanese morphemes) and the syntactic content of the segments defined by various combinations of the following boundary markers: END-

OF-TURN marks, significant pauses, and selected hesitation expressions. The procedures and results of the segment length study are reported in Section 2.

We found that the use of hesitation expressions in various combinations gave segments not much shorter than, or different from, pause units. Accordingly, we will assume that a "first cut" of the speech signal can be made using pauses only, and that most hesitation expressions can be automatically filtered from pause-bounded segments using spotting techniques. Full speech recognition and parsing would begin after this preliminary segmentation and filtering. With these assumptions in mind, we inspected the filtered pause-bounded segments for parsing problems — for structures whose recognition would require extensive adjustments to relatively standard Japanese computational grammars. The results of this survey are described in Section 3.

Attempts to manually translate several dialogues pause unit by pause unit are described in Section 4.

We will conclude with discussion of (1) some parsing experiments related to pause units and (2) a study concerning the relation between pause units and speech acts.

1 Adding Pause Data to Transcripts

This section describes our procedures for augmenting transcripts with accurate data concerning pause placement and duration.

First, marking turn boundaries was not a problem for us. As mentioned, existing transcripts contained the transcriber's judgments concerning turn boundaries. An END-OF-TURN indication was inserted when speaker A stopped talking and speaker B began after a time interval which might approach zero. By contrast, when speaker A talked during B's turn and B continued, a separate notation was used indicating an *aizuchi* or response syllable, comparable to English *um-hmm*. The most frequent Japanese aizuchi in this corpus is *hai*. We accepted the transcriber's judgments in marking turns and in differentiating turns from aizuchi, even though some cases were a bit uncertain: one speaker, for instance, might occasionally start to speak before the other one actually stopped; and there were occasional cases where transcribed aizuchi should perhaps have been noted as turns instead, since they occurred during brief pauses. In any case, except for correcting two or three clear transcription errors, we took turn boundaries as given information, and assumed that automatic turn recognition will not be a major issue for spontaneous speech analysis in the near future. If need be, explicit electronic signals ("push-to-talk" buttons, etc.) can be used to indicate turn boundaries.

Our task was thus to insert into transcribed turns, already separated into tagged morphemes, indications of the locations and lengths of significant pauses. For our purposes, a significant pause was either a juncture of any length where breathing was clearly indicated (sometimes a bit less than 300 milliseconds) or a silence lasting approximately 400 milliseconds or more. If in doubt — for example, if a silence appeared to last 399 milliseconds — we recognized a pause. (This pause length turned out to be long enough to eliminate the silences normally associated with Japanese geminate consonants, as in *chotto*, in which a pause separates the closure and subsequent asperation of a stop consonant. Interestingly, however, there are cases in which the silence is more or less grossly prolonged as a hesitation effect, giving something like *chot ... to*. In such cases, we did recognize pauses for consistency, though these create rather odd segments.)

To facilitate pause tagging, we prepared a customized configuration of the Xwaves speech display program [Xwaves 1993] so that it showed synchronized but separate speech tracks of both parties on screen. The pause tagger, referring to the transcript, could use the mouse to draw labeled lines through the tracks indicating the starts and ends of turns; the starts and ends of segments within turns; and the starts and ends of response syllables which occur during the other speaker's turn. Visual placement of labels was quite clear in most cases (Fig. 1). To check placement, the tagger could listen to the segments delimited by two labels, or could use temporary labels to mark an arbitrary segment for playback. As a secondary job, the tagger inserted a special character into a copy of the transcript text wherever pauses occurred within turns.

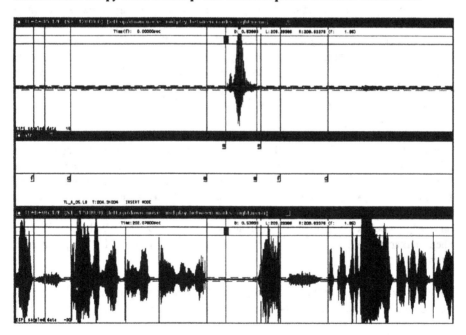

Fig. 1. Interface used by the pause tagger.

After tagging, labels, bearing exact timing information, were downloaded to separate files. Because there should be a one-to-one mapping between labeled pauses within turns and marked pause locations in the transcript, it was then possible to create augmented transcripts by substituting accurate pause length information into the transcripts at marked pause points. Debugging messages were provided in case the matching proved to be inexact, and several errors in the original transcript were found in this way. With practice, the entire tagging and debugging process could be carried out in about 30 tagger minutes per 1 minute of conversation.

Finally, the augmented transcript was transformed into a format similar to that of the ATR Dialog Database [Ehara, Ogura, and Morimoto 1991, 1990], in which each morpheme, hesitation expression, pause, or end-of-turn marker appears on a separate line. (Morphemes are accompanied by identifying codes, and pauses are accompanied by length.) The transformed file could then be used directly as input for programs for (1) calculating segment lengths and (2) pretty-printing the segments defined by a given boundary set for convenience during the syntactic survey.

2 On the Lengths of Segments

In this section we examine the length of segments bounded by pauses and hesitations.

We automatically measured, in Japanese morphemes, the average length and standard deviation of segments marked out by disjunctive sets of boundary markers like (*eeto* END-OF-TURN), where *eeto* is a hesitation syllable comparable to English *uhh*. Two further examples of boundary marker sets are (PAUSE END-OF-TURN) and (END-OF-TURN). A segment terminates whenever any of the listed markers is found in the entire corpus, treated as a long sequence of morphemes. For instance, for the disjunctive set (PAUSE END-OF-TURN), we were measuring how many morphemes one would have to wait, on the average, before encountering either PAUSE or END-OF-TURN.

Zero intervals can occur, as for instance when another marker comes directly before or after END-OF-TURN. For the statistics reported here, we wish to treat a consecutive sequence of markers as a single marker. We thus ignore zero intervals: we remove them from interval lists before averaging. (Since this removal leaves fewer intervals for averaging, an apparent discrepancy between hit count and average interval may be noticed: Assume marker sets A and B yield the same total interval when all intervals are summed, but set A gives a set of segments with more zero intervals than set B does. Then marker set A may give more hits than set B but still yield a longer average interval.)

For each dialogue and for the entire corpus of four dialogues, we performed two sub-studies, which we can call One-by-one and Cumulative.

In the One-by-one study, we assumed turn limits are known, and examined the length of segments which would be marked out by using only one turn-internal marker at a time. That is, each marker set consisted of END-OF-TURN and at most one other marker. In our notation, sample disjunctive marker sets are (PAUSE END-OF-TURN) and (*eeto* END-OF-TURN). We measured number of hits, average segment lengths, and standard deviations for each such set.

The One-by-one study results are summarized in Table 1. Each row gives corpus-wide results for a given marker when paired with END-OF-TURN (with NULL meaning END-OF-TURN alone, with no other marker). The leftmost numbers in a row give summed hits for END-OF-TURN and the marker of interest; but for comparison the hit count for the marker itself (without the hits due to END-OF-TURN) can be seen on the far right. Average interval and standard deviation are also shown for each marker pair. Ordering is top-down from the shortest to the longest average intervals.

	CORPUS TOTALS			W/O END -OF-TURN
	hits	average interval	std. dev	
PAUSE	608	5.89	5.79	205
ano	454	8.32	9.81	51
ne	454	8.49	9.97	51
anou	438	8.68	11.32	35
ee	430	8.95	12.08	27
e	424	9.14	12.55	21
eeto	418	9.18	12.56	15
sou desu ne	408	9.29	12.60	5
etto desu ne	405	9.34	12.93	2
sore de desu ne	406	9.36	12.94	3
eeto desu ne	406	9.36	12.94	3
maa	404	9.37	12.95	1
sore kara desu ne	404	9.37	12.95	1
sou shite desu ne	404	9.37	12.94	1
mazu desu ne	404	9.37	12.95	1
e'	406	9.39	12.96	3
a'	404	9.39	12.96	1
NULL	403	9.39	12.96	0
aa	407	9.41	12.97	4

Table 1. Corpus totals for the One-by-one study

There were 403 occurrences of END-OF-TURN and thus 403 turns in this corpus of 3583 morphemes, i. e. 9.39 morphs per turn on the average, though variation was wide (standard deviation 12.97). There were 205 significant pauses within turns, so about half of the turns contained pauses. If these are used as the only delimiters within turns, the average segment drops to 5.89 morphemes with standard deviation of 5.79. Segments delimited only by significant pauses, in other words, are encouragingly short, roughly 60 percent the length of unsegmented turns. And as mentioned, variability of pause unit length (as shown by standard deviation) is significantly less than that of utterance length.

Significant pauses are several times more frequent than any hesitation expression in this corpus. Hesitation expressions used individually turn out to be ineffective segmentation markers: the most frequent one gives segments only about a morpheme shorter than an average turn. This is because they are often used at the beginnings or ends of turns, and thus are often separated from end-of-turn markers by zero intervals. (While the table lists *ano* and *ne* as tied for first place, the latter is actually the most frequent hesitation expression in this corpus. Several expressions ending with *ne* had to be tabulated separately since they were transcribed as units. Counting all such expressions, *ne* was found 67 times in the corpus. Interestingly, while this particle can be used as a prompt for confirmation, comparable to an English tag question, in this corpus it is almost always used for hesitation instead.)

There was some noticeable individual difference between our two callers, in that Caller2's dialogues contained longer segments than Caller1's dialogues in both telephone-only and multi-media settings. Further, multi-media segments were longer than telephone-only segments for both callers. These effects are indicated in Table 2. Within each column of the table, the format hits/average segment length/standard deviation is used.

We now shift attention to the Cumulative study. In this study our interest was in finding the combination of markers yielding the shortest segments. We tested disjunctive sets of several markers, beginning with END-OF-TURN and adding markers one by one in frequency order to measure the cumulative effect.

Because of the abovementioned effect of zero intervals, the best segmentation combinations do not include all markers. Instead, the shortest average segments are obtained by combining markers which occur medially in turns. The winning combination was (*sou-desu-ne eeto e ee anou ne ano* PAUSE END-OF-TURN). There were 813 hits; average segment length was 5.01; and standard deviation was 4.57.

Even for this best combination, the statistics confirm that hesitation expressions are rather ineffective segmenters, shortening the overall average segment length from 5.89 (standard deviation 5.79) to no less than 5.01 (standard deviation 4.57). That is, the average segment carved out by the winning combination was less than a single morpheme shorter than the segment given by PAUSE and END-OF-TURN.

Once again, the results seemed sensitive to both caller and media. Here we list the "best" combinations (those giving the shortest segments) for each caller-media combination, with statistics once again in the format hits/average segment length/standard deviation: in the telephone setup, Caller1's best boundary set was (*ee anou ne ano* PAUSE END-OF-TURN), yielding 127/4.72/4.51, and Caller2's best set was (*sou-desu-ne eeto e ee anou ne ano* PAUSE END-OF-TURN), giving

265/5.05/4.56; in the multi-media setup, Caller1's best set was (*sou-desu-ne eeto e ee anou ne ano* PAUSE END-OF-TURN), giving 162/5.02/5.03, and Caller2's best was (*etto-desu-ne sou-desu-ne eeto e ee anou ne ano* PAUSE END-OF-TURN), giving 259/5.12/4.28.

| | CALLER1 | | CALLER-2 | |
	TEL	MM	TEL	MM
PAUSE	110/4.96/4.81	185/6.38/6.57	129/5.55/5.52	184/6.19/5.61
ano	86/6.70/7.94	159/7.73/8.59	92/8.27/9.30	117/10.31/12.32
ne	85/6.79/7.97	151/8.43/9.91	94/8.25/9.33	124/9.91/11.43
anou	88/6.52/7.48	141/8.75/10.31	90/8.57/10.29	119/10.29/14.80
ee	90/6.43/7.34	139/9.02/12.06	86/8.92/10.84	115/10.89/15.23
e	85/6.87/8.25	138/9.09/12.00	88/8.90/10.80	113/11.11/16.33
eeto	83/6.90/8.25	136/9.17/12.37	86/8.92/10.76	13/11.11/16.02
sou desu ne	83/6.90/8.25	133/9.34/12.47	85/8.94/10.85	107/11.37/16.03
etto desu ne	84/6.89/8.26	133/9.34/12.47	84/9.05/10.92	107/11.48/16.90
sore de desu ne	83/6.90/8.25	132/9.34/12.56	84/9.05/10.92	106/11.49/16.89
eeto desu ne	83/6.90/8.25	133/9.34/12.53	84/9.05/10.92	106/11.60/16.95
maa	83/6.90/8.25	132/9.34/12.57	84/9.05/10.92	105/11.60/16.96
sore kara desu ne	84/6.80/8.23	131/9.42/12.59	84/9.05/10.92	105/11.60/16.96
sou shite desu ne	83/6.90/8.25	132/9.34/12.54	84/9.05/10.92	105/11.60/16.96
mazu desu ne	83/6.90/8.25	132/9.34/12.56	84/9.05/10.92	105/11.60/16.96
e'	84/6.89/8.26	131/9.42/12.59	85/9.04/10.91	106/11.60/16.96
a'	84/6.89/8.24	131/9.42/12.59	84/9.05/10.92	105/11.60/16.96
NULL	83/6.90/8.25	131/9.42/12.59	84/9.05/10.92	105/11.60/16.96
aa	83/6.90/8.25	131/9.42/12.59	84/9.05/10.92	109/11.68/17.01

Table 2. Corpus totals for the One-by-one study,
showing effects of speaker and media.

3 Syntactic Contents of Pause Units

We next discuss the results of our syntactic survey. To review, we assumed that a "first cut" of the speech signal can be made using pauses only, and that most hesitation expressions can be automatically filtered from pause-bounded segments using spotting techniques. In the resulting filtered, pause-bounded segments, what sorts of syntactic anomalies or problems are encountered beyond those handled in more canonical Japanese grammars? With what frequency do they occur?

To give the main results first, we find that in 608 pause units from which hesitations had been manually filtered, 59 presented syntactic problems that would prevent an internal parse using standard Japanese grammars (as judged manually). Thus a bit more than 90% of the pause units were judged internally parsable.

This figure seems encouraging for a divide-and-conquer analysis strategy. It suggests that pause units are indeed relatively well-formed internally, so that one might indeed separately address (1) the problem of analyzing pause units internally and (2) the problem of fitting analyzed pause units together to form coherent utterances. (An alternative way of describing the division of labor, suggested in conversation by B. Srinivas, is that pause units might provide a partial bracketing of utterances which would be exploited to constrain subsequent analysis.) For the fragment-assembly task, possible techniques include the use of specialized grammars created manually or information-extraction procedures.

Let us now survey the problems seen in the 10% of our pause units which did not appear immediately parsable.

(1) *Clarification paraphrases.* There were numerous examples in which the speaker used a paraphrase to clarify. In such cases, analysis for translation should ideally recognize that a single meaning has been expressed more than once.

In four cases, the speaker immediately followed a referring expression with another, perhaps deictic, expression having the same referent. In one such case, the "agent" said *tsugi no kaidan kochira ga mietekita jiten de*, changing in mid-utterance between "at the roundabout where you saw the next stairs" and "at the roundabout which you saw on this side". She rephrased *tsugi no kaidan* ("the next stairs") with *kochira* ("on this side"), creating real difficulties for both parsing and translation.

In five cases, the clarification paraphrase did not immediately follow the expression to be clarified. Instead, a pause and perhaps an expression signaling clarification or uncertainty intervened. For example, a caller said, *socchi no hou ga omote (pause) tte yuu ka (pause) eki no omote desune*, meaning "The front entrance on that side (pause) that is (pause) the front entrance of the station, right?".

In three cases, the clarification paraphrase was added after the completion of a sentence. For example, a caller said *kore wa futsuu desu ka (pause) futsuu densha?* meaning "Is this a local? (pause) a local train?".

(2) *Correction paraphrases.* In four cases, paraphrase was used following a pause and/or hesitation syllable for correction rather than for clarification. In such repairs, analysis for translation should ideally recognize that a defective expression has been abandoned and replaced by a new one. For example, a caller said, *ano hoteru no tsugou (pause) mada ano tehai shite nai n desu kedo*, meaning "Um the hotel situation (pause) still um the arrangements still haven't been made".

(3) *Fragmentary turns.* In 26 cases, an entire turn contained an incomplete constituent. In general this was because the other party's response intervened. Such responses were usually comprehension signals, and could thus be seen as aizuchi which coincided with pauses.

(4) *Fragments within turns*. There were three cases in which a noun phrase was split by a significant pause. In two additional cases, verb phrases were similarly split. In one of them, a word was cut: the caller said, *wakarimashi (pause) ta* meaning "I under (pause) stand".

(5) *Omission of postpositions*. While some Japanese grammars can handle omission of postpositions *ga, wa, wo*, we also encountered for instance the phrase *basu nottara ii n desu ka* ("Should I ride (on) a bus?"), in which a less usual drop (of the expected *ni*) has occurred. Such omissions seem especially characteristic of spoken style. There were seven postposition dropping examples.

(6) *Questions without* ka. In conversational style, the final question particle *ka* is often omitted, so that intonation and context must be used to recognize the illocutionary act. There were five examples.

We can summarize these results as follows:

clarification paraphrases		12
within sentence, no separation	4	
within sentence, with expression between	5	
after sentence	3	
correction paraphrases		4
fragmentary turns		26
fragments within turns		5
omission of unusual postpositions		7
questions without *ka*		5
--		
Total		59

Table 3. Syntactic difficulties seen in pause units not easily parsable

4 Translation of Pause Units

The final question we address here is whether translation of isolated pause units seems at all possible. To gain an informal impression, we manually translated four dialogues pause unit by pause unit.

Following is a representative sample of the results. We show three turns spoken consecutively by the "agent" in multimodal conversation Caller1-MM. (We ignore Caller1's monosyllabic intervening turns and aizuchi.) We show the turns segmented by pauses, with times in seconds. Hesitations to be filtered are bracketed and not translated. Translations are as literal as possible, and reflect human resolution of ambiguities. In some cases the translation quality is poor, indicating that the segments should be rejoined (1.2 and 1.3) or subdivided by later processing (2.3); in other cases involving the parsing anomalies tabulated above, it may prove impossible to obtain any translation for a pause unit without recourse to special techniques for robust processing.

TURN 1:

1.1   *[/ etto /] / koko / kara / de / shi / ta / ra /* (pause 0.3176)
       "(if it is/if you mean) from here"

1.2   *[/ ano-/]/ kochira* (pause 0.3675)
       "this side"

1.3   */ no / kaidan / wo / aga / t / te / itada / ki / ma / shi / te /* (pause 1.3476)
       "you go up the [modifier missing] stairs and"

1.4   */koko / zutto / wata / t / te / itada / ki / ma / su*
       "you cross all the way here"

TURN 2:

2.1   */de /* (pause 0.3176)
       "and"

2.2   */tsugi / no / kaidan / kochira / ga / mie / te / ki / ta / jiten / de / hidari / ni / maga / t / te / itadaki / ma / su / ichiban / saisho / ni* (pause 0.4038)
       "the next stairs at the roundabout that came into view on this side you turn left first of all"

2.3   */ de / te / ki / ma / su / kono / kousaten / mitai / na / tokoro / wo* (pause 0.3857)
       "you come out. At this place like a crossroads"

2.4   *[/ de / su / ne /] migi / ni / it / te / itadaki / ma / su*
       "you turn right"

TURN 3:

3.1   *de / migi / ni / it / te /* (pause 0.4220)
       "and you go to the right and"

3.2   *itada / i / te / de / koko / no / kaidan / wo / ori / te / itada / ki / ma / su / to /* (pause 0.3811)
       "[missing verb, -te form] for me and when you go down the stairs here"

3.3 *[| ano |] | karasuma | chou | guchi | ni | de | te | mai | ri | ma | su*
"you come out of the Karasumachou exit"

This exercise left us with the impression that while translation by pause units would be far from perfect, it might give understandable results often enough to be useful for certain purposes, even without any attempt to automatically assemble fragments into a coherent whole.

5 Related Work

It is presently common to use separate grammars for speech recognition and linguistic analysis in Japanese, since the first process is phrase (bunsetsu) oriented while the second focuses on the main verb and its complements. It might be possible, however, to use grammars based on a new structural unit to integrate both sorts of processing. [Hosaka, Seligman, and Singer 1994] first suggested the use of pause units in this role (though early parsing experiments were inconclusive); and further studies along these lines have been carried out by [Takezawa et al. 1995].

[Tomokiyo, Seligman, and Fais 1996] present evidence that pause units may be significantly associated with pragmatic units: in this study, 77% of the pause units in a corpus similar to ours coincided with segments defined by surface patterns expressing speech acts. However, the definition of pause units was not identical to ours, since pauses as short as 100 milliseconds were used as boundaries.

Conclusions

We have developed a technique for augmenting transcripts with accurate information regarding pause placement and length. Using this information, we addressed four questions: (1) Are pause units (segments bounded by natural pauses) reliably shorter than utterances? The answer was Yes: on average, pause units were 60% the length of whose utterances, with much less variation. (2) Would hesitation expressions yield shorter units if used as alternate or additional boundaries? The answer was Not much, apparently because pauses and hesitation expressions often coincide. (3) How well-formed are pause units from a syntactic viewpoint? We manually judged that 90% of the pause units in our corpus could be parsed with standard Japanese grammars once hesitation expressions had been filtered from them. (4) Does translation by pause unit deserve further study? The answer was Yes, in that a majority of the pause units in four dialogues gave understandable translations into English when translated by hand.

We are thus encouraged to further study a "divide and conquer" analysis strategy, in which parsing and perhaps translation of pause units is carried out before, or even without, attempts to create coherent analyses of entire utterances. Implementation of this strategy in a working system remains for the future.

Bibliography

Ehara, T., K. Ogura, and T. Morimoto. 1991. "Contents and structure of the ATR bilingual database of spoken dialogues." In *ACH/ALLC*, pages 131-136.

Ehara, T., K. Ogura, and T. Morimoto. 1990. "ATR dialogue database." In *Proceedings of ICSLP*, pages 1093-1096.

Furukawa, R., F. Yato, and K. Loken-Kim. *Analysis of telephone and multimedia dialogues.* Technical Report TR-IT-0020, ATR, Kyoto. (in Japanese)

Hosaka, J. and T. Takezawa. 1992. "Construction of corpus-based syntactic rules for accurate speech recognition." In *Proceedings of COLING 1992*, pages 806-812, Nantes.

Hosaka, J. 1993. *A grammar for Japanese generation in the TUG framework.* Technical Report TR-I-0346, ATR, Kyoto. (in Japanese).

Loken-Kim, K., F. Yato, K. Kurihara, L. Fais, and R. Furukawa. 1993. *EMMI-ATR environment for multi-modal interaction.* Technical Report TR-IT-0018, ATR, Kyoto. (in Japanese).

Morimoto, T., T. Takezawa, F. Yato, et al. 1993. "ATR's speech translation system: ASURA." *Proceedings of Eurospeech-93*, Vol 2., pp. 1291-1294.

Takezawa, T. et al. 1995. *A Japanese grammar for spontaneous speech recognition based on subtrees.* Technical Report TR-IT-0110, ATR, Kyoto.

Tomokiyo, M., M. Seligman, and L. Fais. 1996. "Using Communicative Acts to analyze spoken dialogues." Draft.

Xwaves93. 1993. Entropic Research Laboratory, 1993.

Acknowledgements

The authors wish to thank ATR Interpreting Telecommunications Laboratories for support. We especially thank Kyung Ho Loken-Kim for inspiration and participation during early stages of this research.

Syntactic Procedures for the Detection of Self-Repairs in German Dialogues

Bernd Tischer

Ludwig-Maximilians-Universität München, Germany
Institut für Phonetik und Sprachliche Kommunikation, Schellingstr. 3, D-80799 München
tischer@phonetik.uni-muenchen.de

Abstract. One problem of spoken language processing is the handling of self-interruptions and self-repairs in spontaneous speech. Within a sample of negotiation dialogues and free conversations 4300 self-repairs were collected. The repairs were classified by two kinds of covert repair (hesitations, word repetitions) and four kinds of overt repair (retracings, instant repairs, fresh starts, pivot constructions). Self repairs show syntactic regularities which can be used for automatic processing of spontaneous speech (automatic identification of a repair and automatic transformation into the correct utterance). 96% of the repairs are identified and transformed by eleven detection rules. For only 4% of the repairs the rules cannot be applied. For the detection of these rare cases prosodic cues have to be taken into account.

1 Introduction

One problem of spoken language processing is the handling of self-interruptions and self-repairs in spontaneous speech (see [3], [5], [7] and [8]). Within the domain of speech recognition systems robustness with respect to ungrammatical or incomplete structures is a basic requirement, but most systems fail if the speech input contains self-interruptions, fresh starts or other forms of self-repair. In order to develop a grammar for the processing of spoken language the syntactic and prosodic regularities of self-repairs have to be examined.

Mere self-interruptions are disruptions of the speech flow which are marked by an editing term (*hesitations*, in German e.g., "äh", "ähm", "also") or by *repetition* of words, syllables or single phonemes. After the interruption point the utterance is continued without any repair. However, the interruption indicates that the speaker had some trouble, maybe in solving a content problem, in lexical access or in generating the surface structure. Self-interruptions without overt repair are called *covert repairs* ([1], [4]). In contrast to interruptions without any repair, *overt repairs* contain a repair of the preceding utterance. The main forms are sentence break-offs with a *fresh start* after the interruption (false starts, e.g., "In München ist/- ich brauch fast dreimal so viel wie in Heidelberg"; "I prefer/- perhaps we can find a day in the second week"), *word replacements, inserts, deletions,* and *pivot constructions* ([6]), in which the repair is mediated by a term or a phrase which is both part of the original utterance and the following repair (e.g., "Shall *I* can tell you tomorrow" - the pivot is "I", and the repair is the conversion from a question to an assertion).

If the input of automatic speech processing is spontaneous speech, speech processors should be able to detect repairs and to replace the original utterance by the correct utterance, e.g. for word replacements:

(1) IDENTIFY AS REPAIR: "Es war nicht/- war natürlich nötig"
(It was not/- was of course necessary)

(2) REPLACE "nicht" by "natürlich": "Es war natürlich nötig".
(not)   (of course)   (It was of course necessary)

One aim of the present investigation is the collection of self-repairs and their classification by a system of repair categories. Frequency counts give a first impression of the main forms of self-repair. The second step consists in the formulation of transformation rules by which the original utterance can be transformed into a correct utterance.

A subordinate question of the investigation refers to the effect of situational context on the frequencies and structure of self-repairs. One possible effect is the turn-taking behavior. In natural dialogues, the speaker having trouble must signal that he is continuing his turn (if he does not, he possibly will lose the turn). Therefore, hesitations and/or repetitions of words should increase if there is a possibility for self-selection of the turn by the other speaker. As has been described in [8], in one of the early experimental conditions of VERBMOBIL dialogues (negotiation of appointments) speakers could self-select the turn by speaking even when the other speaker was talking (*uncontrolled turn-taking without button*). In another condition, the speaker had to push a button when he finished his turn. After that, the other speaker could take the turn only by pushing his own speaker button (*controlled turn-taking with button*). Another possible effect is the high amount of problem solving activities in negotiation dialogues, which could raise the probability of self-repairs. To determine the effect of problem solving, the frequencies of repairs within VERBMOBIL dialogues (negotiation of appointments) are compared with the frequencies of repairs in *free conversations* without any problem solving pressure (partners are talking in order to get acquainted).

2 Speech Material

The corpus consists of 8733 turns containing 4300 covert and overt repairs. 3873 turns were drawn from negotiation dialogues without controlled turn-taking (no button pressing), 1390 turns from negotiation dialogues with controlled turn-taking (button pressing), and 3470 turns from free conversation without problem solving (partners getting acquainted). Negotiations without controlled turn-taking contain 1049 instances of repair (27% of all turns). Negotiations with turn-taking control contain 1088 repairs (78% of all turns), and free conversations contain 2163 repairs (62% of all turns). The lower percentage of repairs in negotiations without controlled turn-taking is due to the higher portion of short turns within these dialogues (short answers and back channel responses like "hm"). Negotiations with button pressing (controlled turn-taking) and free conversations contain long turns (e.g., narrations), which raise the probability of making a repair.

2.1 Types of Repair

Classification of all instances of repair (N = 4300 repairs) was done by dividing the repair utterances up into the following types of the general surface structure "OU (original utterance) + DI (disfluency interval) + R (either overt repair or continuation of the OU without any overt repair)":

1. Covert Repairs (62%):

1.1 *Hesitations* with filled pauses "äh" (er) or "hm" within a turn (30%): e.g. "Did you know that you were <er> observed?"

1.2 *Repetitions* of single words or strings of words (32%):
   e.g., "+/the last/+ the last thursday in August"

2. Overt Repairs (34%):

2.1 Sentence break-offs with a fresh start after the interruption (*fresh starts*, 14%): e.g., "I prefer/- perhaps we can find a day in the second week"

2.2 *Replacements* of single words or larger speech segments (14%):
   (a) Immediate continuation of the OU after the DI with the replacing word (*instant repairs*, 8%): e.g., "I have/- was at the university"
   (b) Anticipatory retracing to an item of the OU at the beginning of R (retracings, 6%), e.g., "I have/- I was at the university"

2.3 *Insertion* of one or more words by retracing or instant repair (3%): e.g., "You simply imag-/- you simply can imagine"

2.4 *Deletion* of one or more words of the OU by retracing or instant repair (1%): e.g., "If you don´t want then/- if you want then ..."

2.5 *Pivot constructions* with amalgamation of two kinds of surface structure (2%): e.g., "Shall I can tell you tomorrow".

3. Rest (4%, mainly single phonemes without identifiable target word)

Prototypical instances of overt repairs contain all three utterance parts, *original utterance* (OU), *disfluency interval* (DI) and *repair* (R), in a linear sequence:

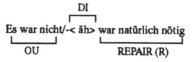

However, in natural dialogues many repairs are produced on the fly. The production of hesitations or editing terms within DI may be useful for repair detection by the listener, but this is no necessary precondition. Counting the overt repairs seperately, in reality only 45% of the instances contain a lexical marked DI (e.g., a filled pause, repetition of a word, word fragments, editing terms like "also", "nein", "gut"). Taking into account that filled pauses, repetitions and word fragments are no necessary conditions for the presence of an overt repair, repair detection should be more successful by looking for syntactical or prosodical cues of self-repair.

Although some studies give evidence that prosodic informations are useful for the detection of repairs if neither lexical nor syntactical repair markers exist ([5]), in this study we are mainly interested in syntactical cues. Because in most instances repairs can be identified by a human researcher reading the transcribed utterances without knowing intonation, loudness, or pause lengths, it is to be supposed that lexical and syntactical informations are the most important cues for repair detection.

Covert repairs. 62% of all repairs (N = 4300) are hesitations (e.g. "äh" without overt repair) or repetitions (mainly word repetitions like "Also +/ich/+ ich hab da keine Zeit"; "/ I/+ I have no time"). Control of turn-taking and problem solving have no influence on the frequency of covert repairs. However, counting the frequencies of hesitations and repetitions seperately, there is a main influence of context:

	Hesitations (filled pauses)	Repetitions
Negotiations (button)	52 %	15 %
Negotiations (no button)	32 %	25 %
Free conversations	17 %	45 %

Table 1. Frequencies of hesitations and repetitions in per cent of all repairs in negotiations and free conversations.

Table 1 shows that the problem solving context in the scheduling task raises the probability of hesitations (filled pauses) whereas uncontrolled turn taking and free conversations raise the probability of word repetitions. It can be concluded that the speaker when having trouble is keeps the floor mainly by word repetitions. On the contrary, nonverbal hesitations (filled pauses) are mainly an outcome of problem solving, particularly if there is no risk of losing the turn.

Overt Repairs. 34% of all repairs are overt repairs. Sentence break-offs with *fresh starts* and *word replacements* within the same sentence are the most frequent types of overt repair. 14% of all repairs are fresh starts (negotiations with button: 10%; without button: 16%; free conversation: 14%). The portion of word replacements is 14% of all repairs (with button: 10%; without button: 12%; free conversation: 16%). Less frequent types of overt repair are word inserts (3%, e.g. "Man +/kann einfach/+ kann *sich* einfach vorstellen), pivot constructions with amalgamation of two kinds of surface structure (2%) and word deletions (1%, e.g. "Also +/wenn *nicht* dann/+ wenn dann würd ich lieber den fünfzehnten nehmen").

3 Syntactic Procedures of Repair Detection and Transformation

In order to identify repairs automatically and to replace the existing structure of the utterance by the correct utterance, the formulation of identification and transformation rules is necessary. For this aim, the surface structure of each repair was described in form of a linear string of word classes. Secondly, the frequencies of the resulting strings were inspected. If there are strings ocurring mainly in repair utterances, syntactical procedures for the detection and transformation of repairs can be formulated. The classification of word classes was done by the grammar of Helbig & Buscha ([2]), e.g.:

(1) Wo das wirklich mal/- und das könnte man gut mischen

(Where that sometimes really/- and that you could really mix)

Rel + Dem + Adv + Adv/- Conjc + Dem + Aux + ProN + Adv + V

(Rel = relative pronoun; Dem = demonstrative pronoun; Conjc = coordinating conjunction; Aux = auxiliary verb; ProN = pronoun; Adv = adverb; V = main verb)

The aim of this analysis is the detection of strings of word classes indicating cases of repair without any verbal disfluency markers (editing terms within DI like "äh", "hm","well") or without prosodic information. In example (1) the word order in the German sentence "und das könnte man gut mischen" is not appropriate to the continuation of the OU "Wo das wirklich mal" because in German the relative pronoun requires a final position of the inflected auxiliary verb "könnte".

3.1 Covert Repairs

All cases of *hesitation* (30% of all repairs) can be identified and eliminated by a lexicon of editing terms (in German "äh", "ähm", "ah" or "hm"). In order to identify *word repetitions*, a segmentation of the speech input at the word class level is necessary because there are instances of repetitions of same words with different word classes, e.g. "die" ("that") as demonstrative pronoun as well as relative pronoun in the sentence "Das sind die, die ich meine" ("That is that that I mean"). 76% of all word repetitions are repetitions of single words, 20% are two-word repetitions (e.g. "Also +/das war/+ das war ganz toll), and 3% are three-word repetitions. In such cases, the OU can be transformed into the correct utterance by a simple deletion rule operating for each immediately repeated item. Although an erroneous deletion of an intensifying word repetition (e.g. "a very very nice day") is possible by this operation, it would change the meaning of the whole utterance only moderately.

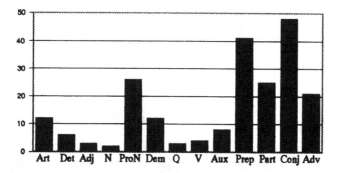

Figure 1. Absolute frequencies of word classes of repeated words in word repetitions based on 211 repeated words from free conversations. Abbreviations: Art (article); Det (determiner, e.g.. "this", "your", "my" plus N); Adj (adjectives and numerals); N (noun); ProN (personal or relative pronoun); Dem (demonstrative pronoun); Q (question pronoun); V (verb); Aux (auxiliar verb); Prep (preposition); Part (particle); Conj (conjunctions); Adv (adverb).

Figure 1 shows, that the main word classes of the repeated items are conjunctions, prepositions, pronouns and particles. On the other hand, the words following the repeated items are content words (lexical heads of NPs and VPs) in most of the cases (see figure 2). Separate countings showed that the probability of word repetitions increases at the beginning of sentence boundaries and at the beginning of larger syntactical constituent (PP, NP, VP). For example, if the first word after a repetition is a pronoun (Figure 2), in 65% of the instances the pronoun is preceded by a repeated sentence conjunction (e.g. "+/and/+ and I went home"). The nouns from figure 2 are preceded by repeated prepositions or by repeated determiners in 85% of the instances.

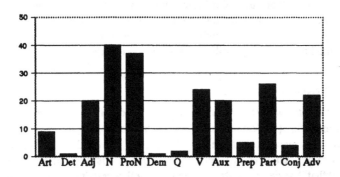

Figure 2. Absolute frequencies of word classes of words following word repetitions. Abbreviations: See Figure 1.

3.2 Replacements

Figure 3 shows that verbs (main verbs, auxiliary verbs and copula) are the most frequently replaced words in overt repairs. In 90% of the instances words were replaced by words of the same word class. Because word replacements include all word classes, the detection of replacements in spoken language processing must be controlled by general detection rules.

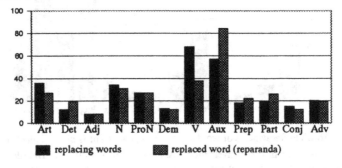

Figure 3. Absolute frequencies of word classes of replacing and replaced words (N=327). Abbreviations: See Figure 1.

Replacements can be identified and transformed by application of two rules ([4]). The *lexical identity rule* is applicable to instances in which the speaker retraces to an earlier word of the OU:

(R1): If the first word r_1 after the interruption of the OU is lexically identical to a word o_i of the OU, then replace o_i by r_1, insert the repair from there on and replace in whatever follows in the OU (e.g., "Ich hab/- ich war schon an der Uni";" I have/ - I was at the university already").

In 99% of all retracings the lexical identity rule is leading to the correct transformation. The only case of incorrect transformation has the following wording:

(2) (OU:) Da sind die *mit* der Sprachproduktion weiter als *mit/-* (DI:) nein (R:) *mit* dem Sprachverständnis weiter.

   (They are *in* speech production better than *in/-* no *in* speech perception better)

In this case, the application of rule (R1) leads to the incorrect transformation

(2´) Da sind die mit der Sprachproduktion weiter als mit dem Sprachverständnis weiter.

   (They are in speech production better than in speech perception better)

The incorrectness is recognizable from the repeating string "weiter" ("better") at the end of (2´). In this instance, also the item r_{1+3} is lexically identical to o_{i+3}. Such errors are prevented by the following expansion of rule (R1):

(R1´) If the first word r_1 after the interruption of the OU is lexically identical to a word o_i of the OU, or if other words r_{1+j} are lexically identical to the words o_i and o_{i+j}, then replace o_i by r_1, insert the repair from there on and replace in whatever follows in the OU.

Another problem is the length of the interval between o_i and r_1. If a recognition system would blindly apply rule (R1´) to all sentences, the probability of false-alarm-transformations would rise if the interval between o_i and r_1 is too large. For example, rule (R1´) would erroneously transform the correct sentence "Da sind die *mit* der Sprachproduktion weiter als *mit* dem Sprachverständnis" ("They are *in* speech production better than *in* speech perception") into the form "Da sind die mit dem Sprachverständnis" ("They are in speech perception") because of the lexical identity of the preposition "mit" ("in"). In order to prevent such false alarm transformations, two results are important. First, in 95% of all retracements there are only one or two words between o_i and r_1. Second, in the remaining 5% of the cases, when there are more than two words between o_i and r_1, speakers retrace to more than one word of the OU at once (e.g., "*Das ist* schon da Richtung/- *das ist* da der ganze Hochschulsport").

The second rule is the *category identity rule*. This rule is applicable to instant repairs, in which the speaker starts immediately with the correct word replacing the problem word in the OU (e.g., "Ich *hab/- war* schon an der Uni"; "I *have/- was* at the university"). The transformation into the correct utterance is possible by the following rule:

(R2) If the word class of the first word r_1 after the interruption of the OU is equal to the last word o_n of the OU, then o_n should be replaced by r_1.

By this rule only 60% of the instant repairs are correctly identified and transformed. However, if the rule is extended to one of the last three words of the OU (o_n, o_{n-1}, o_{n-2}), 94% of all instant repairs can be correctly transformed:

(3a) Ich *hab'* schon/- *hatte* halt (o_{n-1})

    (I *have* yet/- *had* just ...)

(3b) *We* are/- *I* am glad (o_{n-1})

(3c) Meistens *nimmt* man sich/- *nehmen* wir uns was vor (o_{n-2})

    (Mostly one *takes* up/- we *take* in hand something)

(3d) *Is* it in/- *will* she really arrive in time? (o_{n-2})

However, the extension of rule (R2) raises the probability of false alarm recognitions in serial enumerations and specifications, e.g., "*Ich* bin jung, *du* bist alt" ("*I* am young, *you* are old") (o_{n-2}), "Wir haben *kleine* Fehler, *große* Sorgen" ("We have *small* errors, *big* problems") (o_{n-1}). In order to prevent a blind replacement of the first item by another item of the same syntactic category and the same word class, the possible completion of the clause before the first item of the possible R must be checked firstly (examples 3a - 3d show, that the OU is an incomplete clause in cases of "true" replacements). For this aim, rules for identification of false starts (fresh starts) have to be taken into account (see section 3.3, for example rule T4). Other cues are lexical identities between the OU and the repair.

    Word inserts and *deletions* can be handled in the same way, with minimal extensions of the two rules.

3.3 Fresh Starts and Pivot Constructions

The two replacement rules (R1´) and (R2) fail completely in cases of syntax break-offs without replacement and without continuation of the OU, if the speaker produces a fresh start after the marked or unmarked disfluency interval DI, e.g. in German:

(4a) Daß man den Dialekt/- ich kann das nicht so einordnen

    (That you the dialect/- I cannot classify it)

(4b) Daß wirklich die Leute/- also wenn er spricht, dann hört man ihm zu wie ´nem Behinderten

    (That the people really/- if he speaks they listen to him as if he is handicapped)

(4c) Das ist doch/- also da kommen wir vielleicht dem Ganzen ein bißl näher

    (That is surely/- well then we are getting closer a little bit)

(4d) Das war nur/- also eigentlich wollte ich Sprachwissenschaft machen

    (That was only/- well actually I wanted to do linguistics

(4e) Und sind dann/- solche Sätze haben wir ihr dann gesagt

    (And then are/- such sentences we told her then)

(4f) Aber du wirst irgendwann/- das packst du nicht mehr

    (But you will sometimes/- that you don´t lick any longer)

Because 14% of all repairs are fresh starts, and most of them contain neither explicit repair markers nor lexical identities before and after the DI, these cases have to be analysed carefully. Even without any prosodic information, German speakers are able to detect an illformedness in the string of words by simple word order rules, e.g. in the case of (4a):

*(4a) Daß man den Dialekt ich kann das nicht so einordnen

   (That you the dialect I cannot classify it)

In German, the illformedness of this utterance can be detected by the wrong word order (even without knowing the content of the utterance), because the conjunction "daß" requires a final position of the inflected auxiliary verb "kann". Furthermore, the utterance contains two SNPs ("man", "ich") and two ONPs in the accusative case ("den Dialekt", "das"). Dividing the whole utterance up into the OU "Daß man den Dialekt" ("That you the dialect") and into the new sentence "ich kann das nicht so einordnen" ("I cannot classify it") is possible by a search strategy leading to the detection of the starting point of the clause which is governed by the inflected verb "kann" and the main verb "einordnen".

Ignoring possible semantic anomalies of such strings of words, the fresh starts of our German corpus can be described by invariant syntactic structures serving as input for the inductive formulation of general procedures which lead to the detection of the beginning of R and to the deletion of the OU before the first word of R. Both detection of R and deletion of OU is enabled by nine transformation rules leading to a recognition rate of 83% of all fresh starts and pivot constructions. The remaining cases (17% of all fresh starts and pivot constructions, or 2.7 % of all repairs) are detectable only by prosodic cues or by repair markers like word fragments, repetitions or editing terms.

Some examples for transformation rules:

1. *Missing main verb before a sentence conjunction* (works for 7.5% of all cases, e.g. "Weil ich nie/- *obwohl* die Dozenten hier sehr viel besser sind"; "Because I never/- *although* the teachers here are much better"). If a new clause is indicated by a sentence conjunction (e.g., "and", "or", "although") and a following clause, and if the preceding utterance (OU) contains a clause start without an inflected verb before this conjunction (e.g., "Because I never ..."), then the OU is categorized as a false start and will be deleted:

(T1)   IF: SNP (N, ProN, Dem, Rel or Art) plus optional (Part, Adv, Prep or Aux) is located without V before CLUSTER1 (Conj/Det/Q + clause).
   THEN: Delete OU before CLUSTER1

2. *Wrong position of the inflected verb after a subordinating conjunction or relative pronoun* (10.8% of all fresh starts, e.g. "Daß ich das früher oder später/- werde ich das machen"). If in German a new clause is opened by a subordinating conjunction or a relative pronoun followed by a NP or by an adverb, a final position of the inflected verb is required (e.g. "Daß ich das früher oder später machen *werde*"). Otherwise, if the position of the inflected verb is wrong (indicated by the nonfinal position of the inflected auxiliary verb in the string *Aux + NP/PP/Adv*, e.g. "*werde ich* das machen"), the beginning of the subordinating clause of the OU is categorized as a false start and is deleted up to the last word before the start of the clause governed by the main verb and its inflected auxiliay verb (e.g., deletion of "Daß ich das" is leading to the remain-

ing clause "früher oder später werde ich das machen"). Because of the high frequency of sentence constructions with auxiliary verbs, 10.8% of all fresh starts and pivot constructions are detected and transformed by the following rule (the term *Conjs* denotes *subordinating conjunction*):

(T2)   IF: Conjs or Rel plus NP/Adv (plus optional PP/Adv/NP) is located before
       CLUSTER2 (inflected auxiliary (plus optional PP/Adv) plus NP/PP/Adv)
       without V between Conjs/Rel und CLUSTER2.

       THEN: Search first word of the clause of CLUSTER2 and delete OU before this
       clause

3. *Missing completion of a copula by a noun/preposition/adverb* (18%, e.g. "Das ist nicht/- für diese Arbeit mußt Du keine Untersuchungen machen"; "That is not/- for this work you don't have to make an empirical study"). In German as well as in other languages, copula require a completion by a noun phrase, a preposition or an adverb (e.g., "He is a film director/at home/young"). In German, the missing completion is identifiable by an inflected verb, which is either immediately following after the copula or after the copula plus a word X or a phrase X which can be categorized as the start of the clause governed by the inflected verb (e.g., "He is/- a film director was living there"). If the inflected verb is not the beginning of a question with initial position of the verb (e.g., "Das ist nicht für diese Arbeit. *Mußt Du das machen?*" vs. "Das ist nicht/- *für diese Arbeit mußt du das machen*"), the following rule leads to a correct identification of the fresh start:

(T3)   IF: Cop (plus optional Part/Neg/Conj) is located before CLUSTER3 (X +
       inflected verb) without Adv/PP/NP between Cop and CLUSTER3

       THEN: Search first word of the clause of CLUSTER3 and delete OU before this
       clause

4. *Occurence of an inflected auxiliary verb without main verb before another inflected auxiliary verb* (9.3%, e.g. "Also ich *hab'*/- Lebensmittelchemie *hätt'* ich studieren können"; "I *have*/- Chemistry I *should* have studied"). Many fresh starts contain an inflected auxiliary verb in the OU followed by a new construction with another inflected auxiliary verb. Because two inflected auxiliary verbs without a completion of the first one by a main verb indicate two separate clauses, these instances can be identified and transformed by looking for the occurence of two inflected auxiliaries without another verb between them (e.g., "*Do* you/- I *could* pay"):

(T4)   IF: Aux1 (inflected) is located before Aux2 (inflected) without V between Aux1
       and Aux2

       THEN: Search first word of the clause of Aux2 and delete OU before this clause

Other rules refer to the detection of fresh starts by looking for utterances containing the wrong number of conceptual arguments (NPs) of a main verb (e.g., "weil das/- da kriegt man das mit"; "because that/- there you catch that"), by looking for "impossible" strings of word classes like "demonstrative pronoun plus personal pronoun at the sentence start" (only for German, e.g., "Aber *das*/- *wir* müssen auch Seminararbeiten schreiben"), or by deletion of single prepositions, if they are immediately followed by a copula or by a subordinating conjunction. In all instances, recognition of fresh starts is possible without using prosodic information and without looking for vocal repair markers in the disfluency phase (DI).

3.4 Undetectable Cases

Applying the syntactic procedures of repair identification and repair transformation, 17% of the fresh starts and 6% of the replacements by instant repair remain undetectable (corresponding to a total sum of 4% of all overt and covert repairs, or a detection rate of 96%). Looking at the undetectable cases of fresh starts and pivot constructions separately, all utterances can be divided up into two possible single clauses. In those instances only the intonation or an explicit repair marker indicates that the first clause is wrong or incomplete, e.g.:

(5a) In München ist das einfach/- alles ist teuer

(In Munich this is just/- everything is expensive)

(5b) Ich möchte es irgendwie/- <ähm> was die da anbieten, ist unmöglich

(I like it somehow/- what they offer is impossible)

(6a) Aber da ist/- *zu sagen*, ich komm´ von Berlin

(But there is/- *to say* I come from Berlin)

(6b) Was/- *brauchst Du* mehr Geld?

(What/- *do you need* more money?)

In instances like (5a) and (5b), the string of word classes before DI (/-) is a single clause, if the last word is spoken with a falling intonation. Please note that the German word "einfach" (5a) may be used as an adverb (denoting "simple") if it is produced with a falling intonation, or as a graduating particle ("just") if it precedes another adverb (e.g. "it is just wonderful"). However, utterance (5a) is missing the falling intonation, and utterance (5b) includes the repair marker <ähm>. Instances like (6a) and (6b) contain a pivot element (printed in italics) which is a possible completion of the OU. Again, repair detection is possible if a recognition system is using prosodic information as a signal of repair (6a and 6b), or by taking notice of explicit repair markers like the German word "also" ("well"). The most important finding on the supporting function of repair markers is, that undetectable fresh starts and pivot constructions contain significantly more explicit repair markers than the other forms of overt repair (replacements and detectable fresh starts).

4 Conclusion

The present investigation has shown that self repairs produced in German dialogues have invariant syntactic features which can be used for automatic processing of spontaneous speech. 96% of the repairs are identified and transformed correctly by eleven detection procedures. Two procedures (rule R1´ and the modified rule R2) apply to the detection and transformation of replacements, and nine procedures apply to fresh starts and pivot constructions. The input for each of the syntactic procedures is a string of words including only phonological and word class information. However, the rules are not applicable to 4% of the repairs. For the detection of these rare cases prosodic cues have to be taken into account.

Because the results of this descriptive study were obtained by hand segmentation of words and word classes, the basic requirement for repair identification is the recognition of words and word classes within the computed syntactical frame of a linear string of words. Therefore, one aim of further investigations will be the detection of false alarm rates applying the rules to utterances without any repair. Another aim will be the measurement of prosodic features within the three phases OU, DI, and R (intonation, loudness, pauses). Nevertheless, the present findings on repair detection support the assumption that prosodic features are dispensable in the overwhelming majority of self repairs.

References

[1] E.R. Blackmer and J.L Mitton, Theories of monitoring and the timing of repairs in spontaneous speech. *Cognition,* 39, 173-194, (1991).

[2] G. Helbig and J. Buscha, *Deutsche Grammatik. Ein Handbuch für den Ausländerunterricht,* Langenscheidt Verlag Enzyklopädie, Leipzig, 16th edn., 1994.

[3] D. Hindle, Deterministic parsing of syntactic non-fluencies, *Proceedings of the 21st Annual Meeting* (Association for Computational Linguistics, Cambridge, MA), 1983 (pp. 123-128).

[4] W.J.M. Levelt, Monitoring and self-repair in speech, *Cognition,* 14, 41-104, (1983).

[5] Ch. Nakatani and J. Hirschberg, A corpus-based study of repair cues in spontaneous speech. *Journal of the Acoustical Society of America,* 95, 1603-1616, (1994).

[6] E. Schegloff, The relevance of repair to syntax for conversation. In T. Givon (Ed.), *Syntax and semantics. Vol. 12: Discourse and syntax,* Academic Press, New York, 1979 (pp. 261-286).

[7] E. Shriberg, J. Baer and J.Dowding, Automatic detection and correction of repairs in human-computer dialog, *Proceedings of the Speech and Natural Language Workshop* (DARPA, Harriman, NY), 1992 (pp. 419-424).

[8] H.G. Tillman and B. Tischer, Collection and exploitation of spontaneous speech produced in negotiation dialogues, *Proceedings of the ESCA Tutorial and Research Workshop on Spoken Dialogue Systems* (Vigso, Denmanrk, May 30 - June 2, 1995), ESCA and Center for Person-Kommunikation, Aalborg University, 1995 (pp. 217-220).

Utterance Units in Spoken Dialogue*

David R. Traum[1] and Peter A. Heeman[2]

[1] University of Maryland, College Park, MD 20742 USA
[2] University of Rochester, Rochester, NY 14627 USA

Abstract. In order to make spoken dialogue systems more sophisticated, designers need to better understand the conventions that people use in structuring their
speech and in interacting with their fellow conversants. In particular, it is crucial
to discriminate the basic building blocks of dialogue and how they affect the way
people process language. Many researchers have proposed the *utterance unit* as the
primary object of study, but defining exactly what this is has remained a difficult
issue. To shed light on this question, we consider grounding behavior in dialogue,
and examine co-occurrences between turn-initial grounding acts and utterance
unit boundary signals that have been proposed in the literature, namely prosodic
boundary tones and pauses. Preliminary results indicate high correlation between
grounding and boundary tones, with a secondary correlation for longer pauses.
We also consider some of the dialogue processing issues which are impacted by
a definition of utterance unit.

1   Introduction

Current spoken dialogue systems tend to restrict the allowable interaction patterns to
simple exchanges with the user. Unlike natural human conversation, turn-taking is
generally fixed to the expression of a single sentence or speech act, with little flexibility
about when the turn can change. To move beyond this will require a better understanding
of the nature of spoken dialogue. An important starting point is a clear formulation of
the basic units of language production and comprehension. Our opinion is that the best
units of dialogue for an artificial system to attune to are the very same ones that humans
use. Since spoken dialogue systems are meant to interact with humans, they must be
able to use appropriate conventions in responding in a timely and appropriate fashion.

It has often been claimed that for spoken dialogue, *utterances* rather than *sentences*
are the primary object of study [3, 4]. But just what *are* utterances? How are they built
up from more primitive bits of speech, and how do they cohere with other utterances
by both the same and other speakers? Following Bloomfield [2], the term *utterance* has
often been vaguely defined as "an act of speech." The problem is that action comes
in many different types and sizes. As discourse analysis of written text concerns the
relationships between different sentences rather than sentence internal relationships,
discourse analysis of spoken dialogue should concern the relationships between utterances. Finding an appropriate definition of *utterance units* is thus an important starting

* Funding for the second author was gratefully received from NSF under Grant IRI-90-13160
and from ONR/DARPA under Grant N00014-92-J-1512. We would also like to thank James
Allen and the editors of this volume for helpful comments.

point for distinguishing utterance-internal language processes (e.g., phonology, speech repairs) from those that operate at a discourse level, (e.g., turn-taking, grounding, rhetorical relations). We describe some of the needs for a precise utterance unit definition in Section 3.

Analysts have proposed many different definitions of utterances and *utterance units*. The *turn* is the unit of dialogue that has most often been proposed for study as a basic utterance unit. Fries [8], for example, uses the term *utterance unit* to denote those chunks of talk that are marked off by a shift of speaker. Other definitions are based on a variety of information, including syntactic (sentence, clause), semantic/pragmatic (a single proposition or speech act), or prosodic (tunes, stress, silence). We consider some of these in more detail in Section 2. The units we will evaluate specifically are spans of speech terminated by prosodic cues: boundary tones and pauses.

Evaluation of the relative efficacy of utterance units proposals is also difficult. Many of the uses for which one needs such a notion are internal cognitive processes, not directly observable in human conversation. One aspect which *is* observable is the way one participant in a conversation reacts to the speech of another. Much of this reaction will be responsive speech itself, especially if other channels, such as eye gaze, are not available (as in telephone dialogues). Such methods have been used previously in studies of turntaking (e.g., [6, 24]), using the turn-transition as a locus for observation of unit features.

In this study we look not just at the existence of a turn transition, but *whether* and *how* the new speaker reacts to what has been said. According to the view of conversation as a collaborative process, inspired by Clark [4], the conversants are not just making individual contributions, but working together to augment their common ground. This process of *grounding* involves specifically acknowledging the installments of others. By examining which speech is acknowledged, we can have some insight into which installments are viewed as successful, and by correlating these with proposed boundary signals, we can also evaluate the utility of the signals. In Section 4, we discuss the grounding phenomena and the coding scheme we used to mark relatedness, which captures the most immediate aspect of grounding – how a new speaker initially responds to prior speech.

In Section 5, we describe how we used this scheme, along with prosodic markings to label a dialogue corpus, to study this issue, with the results presented in Section 6. We then conclude with some observations on how to extend this study and some implications for spoken dialogue systems.

2 Proposals for Utterance Units

Although there is not a consensus as to what defines an utterance unit, most attempts make use of one or more of the following factors.

- Speech by a single speaker, speaking without interruption by speech of the other, constituting a single *Turn* (e.g. [7, 8, 22, 27].
- Has syntactic and/or semantic completion (e.g. [7, 22, 21, 32]).
- Defines a single speech act (e.g. [22, 16, 20]).

– Is an intonational phrase (e.g. [12, 9, 7, 11]).
– Separated by a pause (e.g. [22, 11, 29, 33]).

While the turn has the great advantage of having easily recognized boundaries,[3] there are several difficulties with treating it as a basic unit of spoken language. First of all, the turn is a multi-party achievement that is not under the control of any one conversant. Since the turn ends only when another conversant speaks, a speaker's turn will have only an indirect relation to any basic units of language production. If the new speaker starts earlier than expected, this may cut off the first speaker in midstream. Likewise, if the new speaker does not come in right away, the first speaker may produce several basic contributions (or units) within the span of a single turn.

From a purely functional point of view, many analysts have also found the turn too large a unit for convenient analysis. Fries, for example, noted that his utterance units could contain multiple sentences. Sinclair and Coulthard [31], found that their basic unit of interaction, the *exchange*, cut across individual turns. Instead, they use *moves* and *acts* as the basic single-speaker components of exchanges. A single turn might consist of several different moves, each of which might be part of different exchanges.

Sacks et. al, [27] present a theory of the organization of turns as composed of smaller *turn-constructional units* (TCUs). At the conclusion of each TCU there occurs a *transition-relevance place* (TRP), at which time it is appropriate for a new speaker to take over (or the current speaker may extend her turn with a subsequent TCU). TCUs thus form a more basic utterance unit, from which turns (and perhaps exchanges) can be built. TCUs have been posited to consist of differing types of syntactic contributions, including lexical, phrasal, clausal, and sentential constructions. Much subsequent work on turn-taking (*e.g.*, [6, 24, 7]) has tried to analyze what features are used to signal a TRP. The features that were examined included syntactic completions, pauses, and various prosodic features including boundary tones.

The difficulty with the syntactic, semantic, and pragmatic categories is that they can be very difficult to segment in spoken dialogue. For instance, should one pick the smallest or largest unit which applies? Very often, speakers add to their current installment, using such cue words as "and", "then", and "because" to link to what has come before. Moreover, speakers involved in spontaneous conversation do not always speak in well-formed units according to these criteria. Many actual productions are either fragmentary, ungrammatical, or both. Sometimes, as in the case of repaired speech, there are *no* correct units as produced, these only emerge from a (sometimes interactive) dialogue process. While it would be interesting to take another look at these candidates, more effort must first be spent on devising reliable rules for coding them consistently in spontaneous spoken dialogue.

We therefore focus here on the prosodic features of speech. When people speak, they tend not to speak in a monotone. Rather, the pitch of their voice, as well as other characteristics, such as speech rate and loudness, varies as they speak. (Pitch is also referred to as the fundamental frequency, or f_0 for short.) Pierrehumbert [25] presented a model of intonation patterns that later was taken as part of the basis for the ToBI (Tones

---

[3] Difficulties still remain, such as how to count turns when more than one conversant is speaking, and in determining whether a particular utterance counts as a backchannel item.

and Break Indices) annotation scheme [30]. This model describes English intonation as a series of highs (H) and lows (L). The lowest level of analysis deals with stressed words, which are marked with either a high or low *pitch accent*. The next level is the *intermediate phrase*, which consists of at least one stressed word, plus a high or low tone at the end of the phrase. This *phrasal tone* controls the pitch contour between the last pitch accent and the end of the phrase. The highest level of analysis is the *intonational phrase* (IP), which is made up of one or more intermediate phrases and ends with an additional high or low tone, the *boundary tone*, which controls how the pitch contour ends.

Another way in which a turn can be segmented is by pauses in the speech stream [29, 33]. Pause-delimited units are attractive since pauses can be detected automatically, and hence can be used to segment speech into chunks that are easier to process with a speech recognizer or parser. It is still controversial, however, whether pause-delimited speech is a good candidate for a definition of an utterance unit. For one thing, pauses can occur anywhere in the speaker's turn, even in the middle of a syntactic constituent. Also, oftentimes the speaker's pitch level will continue at the same level before and after the pause. There is also often some silence around the point of disfluency during a speech repair.

There have also been some more radical proposals for smaller basic units. For example, Poesio [26] proposed that each word or morpheme be a basic unit, accessible to all aspects of the discourse processing, including semantics, reference resolution, and dialogue management. This position has some attractions because there is psycholinguistic evidence that people do in fact use context even at such basic levels and are able to act appropriately (e.g., by gazing at an object) as soon as the necessary words to disambiguate the meaning are heard [34]. Additionally, this position allows a uniform style of processing and representation for phenomena like pronoun resolution which have both intra- and inter-sentential components.

3 Uses for Utterance Units

We feel, however, that such a step as word by word discourse processing goes too far — there is a legitimate role for sub-turn utterance units in distinguishing local from discourse phenomena in dialogue processing. While certain aspects of the local context will be necessary for an agent to attend to, many issues can be better dealt with by attending to the utterance unit and performing some kinds of processing only within UU boundaries, while performing others only between utterances. We briefly review some of these issues in this section.

3.1 Speech Repairs

Heeman and Allen [13] propose that speech repairs should be resolved very early in speech processing. Speech repairs are instances where speakers replace or repeat something they just said, as in this example from [15]:

we'll pick up a tank of uh the tanker of oranges
             reparandum   *alteration*

In the above example the speaker intends that the text marked as the *reparandum* be replaced by the *alteration*. Speakers tend to give strong local clues both to mark the occurrence of speech repairs [18] and to help the hearer determine the intended correction [17].

One type of speech repair of particular interest is the *fresh start*, in which speakers abandon what they are saying, and start over. Exactly how much is abandoned is not explicitly signaled. Consider the following example from the TRAINS corpus [14] (d93-13.3 utt7), with intonation phrase endings marked as '%'.

> two hours %
> um okay %
> so um I guess I 'd like to take um
> and tankers only take orange juice right %

In this example a fresh start occurs starting on the fourth line. From the context, it seems that the speaker is canceling the text starting the word "okay" – this is just the current utterance, not material communicated in previous IPs.

Not all same-turn repairs are speech repairs. Consider the following example (d93-26.2 utt41):

> oh that wouldn't work apparently %
> wait wait %
> let's see %
> maybe it would %
> yeah it would %
> right %
> nineteen hours did you say %

In the above example the speaker changes what he/she is saying, yet it doesn't seem as if a speech repair is taking place. The repair is at a more interactional level, since the speaker needs the participation of the hearer to complete it. For true speech repairs, the processing is much more localized, and perhaps strictly to the current utterance unit. A firm definition for utterance units would thus provide a good metric for deciding which repairs could be done locally, and which repairs need more information and have a more global effect.

3.2 Speech Act Identification

While there is a variety of action that proceeds at different levels throughout spoken dialogue, determination of utterance unit sizes can be helpful for distinguishing certain types of action. As an example, the theory of Sinclair and Coulthard [31] is composed of hierarchical ranks of action, some of which occur across speaker turns, and some of which occur within the turn of a single speaker. The notion of an utterance unit could be useful in isolating which span of speech encapsulates a particular rank of action. Also, the multi-level conversation act theory proposed in [36] used an utterance unit distinction. In that proposal, turn-taking acts occurred within utterance units, grounding acts (see Section 4) occurred at the utterance unit level, and traditional core speech acts were composed of multiple utterance units.

Obviously, for such a theory, the nature of an utterance unit will affect how some acts are classified. For example, the set of grounding acts from [36] included acts for `repair` and `continue` functions. Trivially, smaller utterance units would lead to more continue acts, but consider the case in which a repair (e.g., a speech repair) occurs *within* an utterance unit. In this case, the function of the entire unit with respect to the previous context would be considered, and the unit might be seen as performing a continue function. If, however, such a unit were split up so that the alteration was in one unit and the reparandum is in a previous unit, the new unit might be termed a repair.

While this might not pose a serious problem for processing within a system, it does make it much harder to judge validity of such a coding scheme. Although there is a general agreement that it is important for dialogue systems to attend to action, there is still not much agreement on what the proper set of actions should be. For action types which are sensitive to an utterance unit distinction, performing any evaluation of the consistency of action marking becomes next to impossible if the proper set of units are not held constant for different coders.

3.3 Dialogue Management

There are a number of ways in which a good definition of utterance units would facilitate the dialogue management tasks of a spoken NL dialogue system. For one thing, the basic decision of when and how to respond to input from a user can be informed by awareness of utterance units. Even if full discourse processing of language proceeds within an utterance unit, as suggested by [26], the decision of when to respond still mainly follows regular patterns, occurring at regular places [27]. Awareness of the markers for turn-endings and TRPs can help with the decision of when to respond and when to wait for further input.

More importantly, the nature of the previous utterance unit can be a good signal of *how* to respond. Suppose, for instance, the speaker projects a desire to continue speaking. In this case, unless there is some major problem with the previous input, the system can issue a backchannel type response, without detailed processing (yet) of the implications of the speaker's contribution.

On the other hand, suppose there is some problem in understanding the current input. In this case, there are two main options left to the system: (1) wait for the user to say more and hopefully clarify or repair the problem, (2) notify the user of the problems, requesting some sort of repair. Attending to utterance unit production can help in deciding between these two options – in the midst of a unit, it will be preferable to wait for the user. With any luck, the speaker will decide to repair the problem, in which case the speech repair techniques proposed by [13] will fix the problem. At an utterance unit boundary, the dialogue manager can shift strategies, dependent on the anticipated continuation by the user. Given a signal that the user would like to continue her turn, if the problem is some sort of ambiguity or under-specification that is likely to be cleared up by future speech, the system can continue to wait. If, however, the problem is an error in understanding (or disagreement about) the previous utterance, the system can decide to break in with a repair.

An atunement to utterance units can also help a dialogue manager decide how to structure the speech output of the system. By using regular utterance units, a system could provide the same resources to a user for providing feedback of understanding. Also, a system could structure the output in order to request help from the user, such as by ending an utterance unit at a point where the user could fill in a referring expression that the system is having difficulty calculating. In general, the division of contributions into more manageable sized utterance units, which can build up turns of speech, will be a more efficient mechanism for flexibly coordinating the contributions of the conversants to reach a mutual understanding of what has been said.

4   Grounding and Relatedness

As our method for checking the adequacy of proposed utterance unit boundaries, we consider the phenomenon of *grounding*: the process of adding to common ground between conversants [5]. Clark and Schaefer present a model of grounding in conversation, in which *contributions* are composed of two phases, *presentations* and *acceptances*. In the presentation phase, the first speaker specifies the content of his contribution and the partners try to register that content. In the acceptance phase, the contributor and partners try to reach the *grounding criterion*: "the contributor and the partners mutually believe that the partners have understood what the contributor meant to a criterion sufficient for the current purpose." Clark and Schaefer describe several different methods that are used by the partners to accept the presentation of the contributor. These include feedback words such as *ok*, *right*, and *mm-hm*, repetition of the previous content, and initiation of a next relevant contribution. Completions and repairs of presentations of the contributor also play a role in the grounding process.

Traum and Allen [35] built on this work, presenting a *speech acts* approach to grounding, in which utterances are seen as actions affecting the state of grounding of contributed material. In addition to acts which present new material, there are *acknowledgment* acts which signal that the current speaker has understood previous material presented by the other speaker, as well as *repairs* and *requests for repair*. Acknowledgment acts include three types, *explicit* acknowledgments which are one of the feedback words, whether they appeared as a backchannel or not, *paraphrases* of material presented by the other speaker, and *implicit* acknowledgments, which display understanding through conditional relevance. Explicit acknowledgments have also been studied by Novick and Sutton [23], who catalogued the interaction sequences that they play a part in.

4.1   Relatedness

For the purposes of determining an appropriate utterance unit boundary, we are not as concerned with whether something is ultimately grounded, but rather whether the responding agent *starts* the grounding process (with a repair or some component of the acceptance phase), and *which* previous utterances are being grounded. We thus use a less detailed labeling of utterances than that of [35], with a slightly different focus. We lump the repair, request for repair, and the paraphrase and implicit categories of acknowledgment together into one category we call *related*. While, for a particular utterance, it can be difficult to judge *which* one of of these functions is being performed, it is usually straightforward to determine whether or not it performs one of them. We also separate out the *explicit* acknowledgments, since, while they generally perform some sort of acknowledgment, it is not possible to tell with certainty *what* they are acknowledging. Likewise, for utterances that follow backchannels and other turns that consist solely of these signals, there is no real content for the new turn to be related *to*. The third major category is for *unrelated* utterances, which either introduce new topics, or cohere only with previous speech by the same speaker and do not play a grounding role towards presentations by the other speaker. Our final category is for those utterances for which it is *uncertain* whether they are related or not.

4.2 Relatedness Coding Scheme

We used single character annotations to indicate relatedness of utterances. These characters are shown on the line above the text that they label. The coding scheme is explained below, including some examples.

Category: Explicit Ack
Label: e
Description: This category includes all utterances by a new speaker that begin with an explicit acknowledgment – the first word of the turn by the new speaker is a word like "okay", "right", "yeah", "well", or "mm-hm" (also "no" and "nope" in a negative polarity context). This category is also used for repair requests for which it is not clear exactly what should be repaired (e.g., an utterance consisting solely of "what?").
Example: d93-18.3 utt11-12

```
u: so we have to start in Avon
      e
s: okay
```

Category: Related
Description: This category is for those utterances that *demonstrate* a relationship to a previous utterance. This can be a demonstration style acknowledgment (i.e., repetition or paraphrase of a previous utterance by the other speaker), a repair or repair request, or some kind of continuation. Any second part to an *adjacency pair* [28] (such as an answer to a question, or the acceptance or rejection of a proposal) fits in this category. In addition, the recency of the relationship is marked as follows:
Subcategories:

Label	Description
0	utterances **related to the most recent** UU by the previous speaker.
1	related to UU one **previous** to most recent but *not* related to most recent.
2	related to UU two previous to most recent but nothing more recent.
etc.	higher numbers for related to utterances further back.
,	related to previous material by the other speaker, but it is **unclear** to the coder which unit they are related to.

Category: Unrelated

Label: u

Description: these utterances are **unrelated** to previous speech by the previous speaker. This could be either the introduction of a new topic, or could show a relation to previous speech by the same speaker, while ignoring intervening speech by the other speaker.

Example: d93-18.3 utt13-18

```
U: how long does it take to bring engine one to Dansville
   0
S: three hours
   e
U: okay <sil> and then <sil> back to Avon to get the bananas
   0
S: three more hours si(x)- six in all
   u
U: how long does it take to load the bananas
```

Category: Uncertain

Label: ?

Description: The coder is **uncertain** whether or not these utterances relate to previous speech by the other speaker.

Combination Category: After Explicit

Label: X-e

Description: A suffix of -e is used for those utterances which are **after** turns which consist *only* of **explicit** acknowledgments (or backchannels). Since these utterances do not have any content that new material *could* be related to, the categories for unrelated to last (**u**, **1**, etc.), are not very significant. Therefore, we append the marking -e to the other category in these cases.

Subcategories:

Label	Description
u-e	material which is unrelated to previous material by other speaker, but for which the last UU by the other speaker was an explicit acknowledgment.
1-e	material which is related to the penultimate utterance unit by the other speaker, but for which the last utterance unit by the other speaker contained just an explicit acknowledgment. This signal is often used for popping sub-dialogues, such as in the following example, in which the last turn by speaker S refers to the first utterance by speaker U.

Example: d93-18.3: utt51-54

```
U: how long does that take <sil> again
   0
S: that ta(ke)- <sil> so just to go from Bath to Corning
   e
U: mm-hm
   1-e
S: two hours
```

5 Data

As part of the TRAINS project [1], a corpus of problem-solving dialogs has been collected in which a railroad system manager (labeled U) collaborates with a planning assistant (labeled S) to accomplish some task involving manufacturing and shipping goods in a railroad freight system. Since this corpus contains dialogues in which the conversants work together in solving the task, it provides natural examples of dialogue usage, including a number of tough issues that computer systems will need to handle in order to engage in natural dialogue with a user. For instance, our corpus contains instances of overlapping speech, backchannel responses, and turn-taking: phenomena that do not occur in collections of single speaker utterances, such as ATIS [19]. For our current study, we looked at 26 dialogues from the TRAINS-93 corpus [14]. This totaled over 6000 seconds of spoken dialogue comprising 1366 total turn transitions.

5.1 Prosodic markings

All turn-initial utterances are marked with respect to their direct relatedness to the previous full or partial intonation phrase (IP) by the previous speaker. Full IP's are terminated by a boundary tone (labeled %). Partial phrases are anything from the last complete IP to the beginning of speech by the new speaker. If there was no previous IP by the previous speaker than the entire turn so far is counted as a partial IP. The amount of silence between turns is also noted, labeled below in seconds within square brackets (e.g., [.42]).

5.2 Relatedness Markings

Each turn-transition was marked as to how the initial installment of the new turn related to the last few completed or uncompleted IPs produced by the other speaker, using the coding scheme described in Section 4.2. For cases of overlapping speech, the speech of the new speaker was marked for how it relates to the current ongoing installment by the continuing speaker. For simultaneous starts of IPs, only the speech by the new speaker was marked for how it related to the previous speech by the current speaker. The continuing speech of one speaker was not marked for how it related to embedded speech by the other speaker, including true backchannels,

Table 1 summarizes this coding scheme described in Section 4.2, as used to mark turn-initial relatedness to utterance units from the previous turn.

5.3 Example

Below we show some examples of labeled dialogue fragments. The prosodic and relatedness markings are shown above the line they correspond to. The first example shows a simple sequence of 0 and e relations, all following boundary tones, with clean transitions. The first response by S shows a relationship to the first contribution by U. Then the last two start with explicit acknowledgments.

Label	Description
e	explicit acknowledgment (e.g., "okay", "right", "yeah", "well", or "mm-hm")
0	related to the most recent utterance by the previous speaker
1	related to the UU one previous to the most recent but not to the most recent
2	related to utterance two previous to the last one (and not to anything more recent)
,	related to previous material by the other speaker, but it is unclear to the coder whether they are related to the immediately previous UU (which would be marked 0), or to an UU further back (which would be marked 1, or 2, etc.)
u	unrelated to previous speech by the old speaker
?	uncertain whether these utterances relate to previous speech by the other speaker
u-e 1-e	the same meaning for the first item, but follows a turn by the other speaker consisting only of an item marked e

Table 1. Relatedness Markings

Example: d93-13.2: utt18-22

```
                                        % [.42]
U: how long is it from Elmira to Dansville
   0                              % [1.23]
S: Elmira to Dansville is three hours
   e  %
U: okay um so why don't uh
                                        % [1.42]
   I send engine two with two boxcars to Corning
   e  %
S: okay
```

6 Results

6.1 Prevalence of Grounding Behavior

Tabulating the markings on the beginning of each turn yields the results shown in Table 2. This shows how the next utterance is related to what the other speaker has previously said, and so gives statistics about how much grounding is going on. Of all turns, 51% start with an explicit acknowledgment (category e); 29% are related to previous speech of the other speaker (one of the following categories: 0 1 2 , 1-e 2-e ,-e); 15% are unrelated to previous speech of the other speaker, but follow an acknowledgment (u-e); 2% are possibly related or possibly unrelated (category ?), and only 3% are clearly unrelated and do not follow an acknowledgment.

These results give strong evidence of grounding behavior at turn transitions. Fully 80% of utterance display grounding behavior, while another 15% occur in positions in which (according to the theory of [35, 36]) further grounding is unnecessary. It is only in 3-5% of turn transitions in which a lack of orientation to the contributions of the other speaker is displayed.

Category	#	%
Explicit	696	51%
Related	400	29%
Unrelated after Explicit	199	15%
Unrelated	42	3%
Uncertain	29	2%
Total	1366	100%

Table 2. Prevalence of Grounding Behavior

6.2 Boundary Tones

Table 3 shows how relatedness correlates with the presence of a boundary tone on the end of the preceding speech of the other speaker. Here, we have subdivided all of the markings into two groups, those that occur at a smooth transition between speaker turns (*clean transitions*), and those in which the subsequent speech overlaps the previous speech (*overlap*). For the overlap cases, we looked for a boundary tone on the last complete word before the overlapping speech occurred. The distribution of the overlaps into tone and no-tone categories is still somewhat problematic, due to the potential projectability of IP boundaries [10]: a new speaker may judge that the end of a unit is coming up and merely anticipate (perhaps incorrectly) the occurrence of a tone. Thus for some of the entries in the second to last column, there is a boundary tone which occurs after the onset of the new speaker's speech.

Type	Clean Transitions			Overlaps		
	Tone	No Tone	% Tone	Tone	No Tone	% Tone
e	501	24	95%	77	94	45%
0	267	17	95%	16	41	28%
1,2	7	4	64%	7	11	39%
,	9	4	69%	1	6	14%
1,2-e	7	0	100%	3	0	100%
u	18	7	72%	2	15	12%
u-e	186	2	99%	5	6	45%
?	17	3	85%	6	3	67%
Total	1012	61	94%	117	176	40%

Table 3. Boundary Tones and Relatedness

For the clean transitions, we see that more than 94% of them follow a boundary tone. Of more interest is the correlation between the relatedness markings and the presence of a boundary tone. For explicit acknowledgments and utterances that are related to the last utterance, we see that 95% of them follow a boundary tone. For transitions in which the next utterance relates to an utterance prior to the last utterance, or is simply unrelated, we see that only 64% and 72% of them, respectively, follow a boundary tone. This difference between related to last (0) and related to prior and unrelated (1, 2, and u) is statistically significant (p=0.0016).

6.3 Silences

We next looked at how the length of silence between speaker turns (for clean transitions) correlates with boundary tones and relatedness markings. The relatedness markings that we looked at were related-to-last (0), and unrelated-to-last (1 2 u). Due to the sparseness of data, we clustered silences into two groups, silences less than a half a second in length (labeled *short*), and silences longer than a half a second (*long*). The results are given in Table 4.

Type	Tone			No Tone		
	Short	Long	% Long	Short	Long	% Long
0	160	107	40%	6	11	65%
u,1,2	15	10	40%	8	3	27%

Table 4. Silences

We find that when there is a boundary tone that precedes the new utterance, there is no correlation between relatedness and length of silence (a weighted t-test found the difference in distributions for related-to-last and unrelated-to-last to not be significant, with p=0.851). This suggests that the boundary tone is sufficient as an utterance unit marker and its presence makes the amount of silence unimportant.

In the case where there is no boundary tone, we see that there *is* a correlation between length of silence and relatedness markings. Only 27% of unrelated transitions follow a long pause (the mean silence length was 0.421 seconds, with a standard deviation of 0.411), while 65% of the related transitions follow a long pause (the mean silence length was 1.072 seconds, with a standard deviation of 0.746). Although there are few data points in these two categories, a weighted means t-test found the difference in the distributions to be significant (p=0.014). Thus, long pauses are positively correlated with the previous utterance being grounded. So, if the hearer wishes to reply to speech not ending with a boundary tone, he is more likely to wait longer (perhaps for a completion or repair by the speaker) than otherwise. Thus, silences seem to be a secondary indicator of utterance unit completion, important only in the absence of a boundary tone.

7 Discussion

Our results are still preliminary, due to the small sample of the relevant categories. However, they do show several things convincingly. First, although grounding behavior is prevalent throughout these problem solving dialogues, there are different degrees to which the speech is grounded. Since adding to the common ground is a prime purpose of conversation, grounding should prove a useful tool for further investigating utterance units and other dialogue phenomena. Second, the claim that utterance units are at least partially defined by the presence of an intonational boundary seems well supported by the conversants' grounding behavior: in addition to serving as a signal for turn-taking, boundary tones also play a role in guiding grounding behavior. Finally, the grounding behavior suggests that pauses play a role mostly in the absence of boundary tones.

There are several ways in which to follow-up this study. First would be just to examine more dialogues and thus get a larger sample of the crucial *related to prior* category, which indicates that the last (partial) unit was deficient but a previous unit was worth grounding. Another possibility would be to look more closely at the overlapped category to gain a handle on projectability, and look closely at the contrast in distributions from the clean transition case. Even our preliminary numbers from Table 3 show that the overlap category contains more cases of no boundary tone, which is what we would expect in a case of disagreement over turn-taking. Finally, it would be very interesting to apply this analysis to other corpora which occur in other domains (which might have different implications with respect to the grounding criterion), other languages (which may differ in how they signal UU completion), and multi-modal communication, in which conversants have other ways to communicate, as well as speech.

7.1 Implications for Spoken Language Systems

In Section 3, we described some of the benefits for spoken language systems that could be derived from having a clear notion of utterance units. The results of this study show that the use of prosodic information will play a key role in the ability to make use of such units. Recognition of prosodic features, especially boundary tones will be crucial to the determination of utterance units. Also, production of appropriate prosodic features will be essential to achieving useful contributions. Since silence can be seen as the termination of an utterance unit in the absence of a boundary tone, it is also important for a production system to avoid pausing where it would not like a user to make such a judgment of unit finality. This could be achieved by using a continuation tune, or perhaps a filled pause.

One of the important features of utterance units is due to the role they play in grounding. They seem to be the basic unit by which human speakers coordinate to ground their contributions. Hence, utterance unit boundaries define appropriate places for the machine to offer backchannel responses, or to check its space of interpretations to determine whether it is appropriate to give positive (e.g., an acknowledgment) or negative (e.g., a repair) feedback.

References

1. James. F. Allen, L. K. Schubert, G. Ferguson, P. Heeman, C. H. Hwang, T. Kato, M. Light, N. Martin, B. Miller, M. Poesio, and D. R. Traum, 'The TRAINS project: a case study in building a conversational planning agent', *Journal of Experimental and Theoretical Artificial Intelligence*, 7, 7–48, (1995).

2. Leonard Bloomfield, 'A set of postulates for the science of language', *Language*, 2, 153–164, (1926).

3. Gillian Brown and George Yule, *Discourse Analysis*, Cambridge University Press, 1983.

4. Herbert H. Clark, *Arenas of Language Use*, University of Chicago Press, 1992.

5. Herbert H. Clark and Edward F. Schaefer, 'Contributing to discourse', *Cognitive Science*, 13, 259–294, (1989). Also appears as Chapter 5 in [4].

6. Starkey Duncan, Jr. and George Niederehe, 'On signalling that it's your turn to speak', *Journal of Experimental Social Psychology*, 10, 234–47, (1974).

7. Cecelia Ford and Sandra Thompson, 'On projectability in conversation: Grammar, intonation, and semantics'. Presented at the *Second International Cognitive Linguistics Association Conference*, August 1991.

8. Charles Carpenter Fries, *The structure of English; an introduction to the construction of English sentences.*, Harcourt, Brace, 1952.

9. James Paul Gee and Francois Grosjean, 'Saying what you mean in dialogue: A study in conceptual and semantic co-ordination', *Cognitive Psychology*, 15(3), 411–458, (1983).

10. Francois Grosjean, 'How long is the sentence? Predicting and prosody in the on-line processing of language', *Linguistics*, 21(3), 501–529, (1983).

11. Derek Gross, James Allen, and David Traum, 'The Trains 91 dialogues', Trains Technical Note 92-1, Department of Computer Science, University of Rochester, (June 1993).

12. M. A. Halliday, 'Notes on transitivity and theme in English: Part 2', *Journal of Linguistics*, 3, 199–244, (1967).

13. Peter Heeman and James Allen, 'Detecting and correcting speech repairs', in *Proceedings of the 32th Annual Meeting of the Association for Computational Linguistics*, pp. 295–302, Las Cruces, New Mexico, (June 1994).

14. Peter A. Heeman and James F. Allen, 'The Trains spoken dialog corpus', CD-ROM, Linguistics Data Consortium, (April 1995).

15. Peter A. Heeman, Kyung-ho Loken-Kim, and James F. Allen, 'Combining the detection and correction of speech repairs', in *Proceedings of the 4rd International Conference on Spoken Language Processing (ICSLP-96)*, pp. 358–361, Philadephia, (October 1996). Also appears in *International Symposium on Spoken Dialogue*, 1996, pages 133-136.

16. Alon Lavie, Donna Gates, Noah Coccaro, and Lori Levin, 'Input segmentation of spontaneous speech in JANUS: a speech-to-speech translation system', in *Dialogue Processing in Spoken Language Systems*, eds., Elisabeth Maier, Marion Mast, and Susann LuperFoy, Lecture Notes in Artificial Intelligence, Springer-Verlag, Heidelberg, (1997). In this volume.

17. Willem J. M. Levelt, 'Monitoring and self-repair in speech', *Cognition*, 14, 41–104, (1983).

18. R. J. Lickley and E. G. Bard, 'Processing disfluent speech: Recognizing disfluency before lexical access', in *Proceedings of the 2nd International Conference on Spoken Language Processing (ICSLP-92)*, pp. 935–938, (October 1992).

19. MADCOW, 'Multi-site data collection for a spoken language corpus', in *Proceedings of the DARPA Workshop on Speech and Natural Language Processing*, pp. 7–14, (1992).

20. M. Mast, R. Kompe, S. Harbeck, A. Kießling, H. Niemann, E. Nöth, E. G. Schukat-Talamazzini, and V. Warnke, 'Dialog act classification with the help of prosody', in *Proceedings of the 4rd International Conference on Spoken Language Processing (ICSLP-96)*, Philadephia, (October 1996).

21. M. Meteer and R. Iyer, 'Modeling conversational speech for speech recognition', in *Proceedings of the Conference on Emphirical Methods in Natural Language Processing*, Philadephia, (May 1996).

22. Shin'ya Nakajima and James F. Allen, 'A study on prosody and discourse structure in cooperative dialogues', *Phonetica*, 50(3), 197–210, (1993).

23. David Novick and Stephen Sutton, 'An empirical model of acknowledgement for spoken-language systems', in *Proceedings ACL-94*, pp. 96–101, Las Cruces, New Mexico, (June 1994).

24. Bengt Orestrom, *Turn-Taking in English Conversation*, Lund Studies in English: Number 66, CWK Gleerup, 1983.

25. J. B. Pierrehumbert, 'The phonology and phonetics of english intonation', Doctoral dissertation, Massachusetts Institute of Technology, (1980).

26. Massimo Poesio, 'A model of conversation processing based on micro conversational events.', in *Proceedings of the Annual Meeting of the Cognitive Science Society*, (1995).

27. H. Sacks, E. A. Schegloff, and G. Jefferson, 'A simplest systematics for the organization of turn-taking for conversation', *Language*, 50, 696–735, (1974).

28. Emmanuel A. Schegloff and H. Sacks, 'Opening up closings', *Semiotica*, 7, 289–327, (1973).

29. Mark Seligman, Junko Hosaka, and Harald Singer, '"pause units" and analysis of spontaneous japanese dialogues: Preliminary studies', in *Dialogue Processing in Spoken Language Systems*, eds., Elisabeth Maier, Marion Mast, and Susann LuperFoy, Lecture Notes in Artificial Intelligence, Springer-Verlag, Heidelberg, (1997). In this volume.

30. K. Silverman, M. Beckman, J. Pitrelli, M. Ostendorf, C. Wightman, P. Price, J. Pierrehumbert, and J. Hirschberg, 'ToBI: A standard for labelling English prosody', in *Proceedings of the 2nd International Conference on Spoken Language Processing (ICSLP-92)*, pp. 867–870, (1992).

31. J. M. Sinclair and R. M. Coulthard, *Towards an analysis of Discourse: The English used by teachers and pupils.*, Oxford University Press, 1975.

32. Andreas Stolcke and Elizabeth Shriberg, 'Automatic linguistic segmentation of conversational speech', in *Proceedings of the 4rd International Conference on Spoken Language Processing (ICSLP-96)*, (October 1996).

33. Kazuyuki Takagi and Shuichi Itahashi, 'Segmentation of spoken dialogue by interjection, disfluent utterances and pauses', in *Proceedings of the 4rd International Conference on Spoken Language Processing (ICSLP-96)*, pp. 693–697, Philadephia, (October 1996).

34. M. K. Tanenhaus, M. J. Spivey-Knowlton, K.M. Eberhard, and J. C. Sedivy, 'Integration of visual and linguistic information in spoken language comprehension.', *Science*, 268(3), 1632–34, (1995).

35. David R. Traum and James F. Allen, 'A speech acts approach to grounding in conversation', in *Proceedings 2nd International Conference on Spoken Language Processing (ICSLP-92)*, pp. 137–40, (October 1992).

36. David R. Traum and Elizabeth A. Hinkelman, 'Conversation acts in task-oriented spoken dialogue', *Computational Intelligence*, 8(3), 575–599, (1992). Special Issue on Non-literal language.

Development Principles for Dialog-Based Interfaces

Alicia Abella[1], Michael K. Brown[2] and Bruce Buntschuh[1]

[1] AT&T Laboratories, 600 Mountain Ave., Murray Hill,NJ 07974,USA
[2] Bell Laboratories 600 Mountain Ave., Murray Hill,NJ 07974,USA

Abstract. If computer systems are to become the information-seeking, task-executing, problem-solving agents we want them to be then they must be able to communicate as effectively with humans as humans do with each other. It is thus desirable to develop computer systems that can also communicate with humans via spoken natural language dialog. The motivation for our work is to embody large applications with dialog capabilities. To do so we have adopted an object-oriented approach to dialog.

In this paper we describe a set of dialog principles we have developed that will provide a computer system with the capacity to conduct a dialog. Our ultimate goal is to make the job of an application developer interested in designing a dialog-based application to do so effectively and easily.

1 Introduction

In this paper we describe a set of dialog principles that we have developed that will provide a computer system with the capacity to conduct a dialog with a human. It was our goal to choose a set of implementable dialog principles paying particular attention to its applicability. These dialog principles form the basis of our dialog manager.

In general dialog systems fall into one of several categories: *Question-Answer Systems, Spoken Input System,* and *Variable Initiative Systems*[3] . Several dialog systems have been designed for domains where the user seeks a particular piece of information that the system possesses (*Question-Answer Systems*). Some sample systems include Carberry's [5] question-answer systems, IREPS, which focuses on plan recognition and user model usage. Frederking's [7] interest is in handling elliptical utterances as is Hendrix's [9]. Bolc [3] describes the deduction process in answering questions. Raskutti [10] presents a mechanism for generating disambiguating queries during information-seeking dialogues as does Cohen [6] for advice-giving systems. Wilensky developed a question-answer system for answering questions about the UNIX operating system [13]. Grosz [8] developed a system called TEAM that was built to test the feasibility of building a natural language system that could be adapted to interface with new databases. Another system built for providing natural language access for database systems is Hendrix's LADDER [9].

---

[3] We use the names given in [12] for these categories.

What all of the aforementioned systems have in common is their input modality – keyboard input. The *spoken input systems* have the added difficulty of requiring very robust natural language understanding capabilities. One such system is the MINDS system [15]. It uses discourse and dialog knowledge to aid the speech recognition task. Another spoken input system is the TINA system [11]. This system uses probabilistic networks to parse token sequences provided by the speech recognition system.

Another spoken dialog system is VODIS [14], a voice operated database inquiry system for a train timetable application. This work focuses on the use of an integrated framework for dealing with interface design, how dialog can aid the recognition process, and the requirements on dialog control needed for generating natural speech output. Their approach to dialog design is an object-oriented one. Like us, they believe that the generalizations afforded by the inheritance mechanisms of an object-oriented approach will facilitate the construction, augmentation and testing of individual system components. Work on another train timetable application can be found in [1]. Their goal is the design of a system that is user friendly, meaning that they are allowed to talk to the system in an unconstrained and natural manner to extract train information.

Variable initiative systems are those systems capable of taking the initiative when need be but also of knowing when to relinquish the initiative if it determines that the user's input will help guide it to a quicker solution. An example of such a system is Bobrow's GUS [2].

For our dialog manager we have identified a set of general dialog principles. We then provided our system with a set of objects to handle each of these dialog principles. The generality of these principles facilitates the applicability of our system to a variety of tasks.

Our approach to dialog management is object-oriented and rule-based. Unlike a finite-state or directed graph approach, our approach makes the application developer's job easier for large applications because it is scalable. This means that information about each part of the application is encapsulated, and hence easy to augment, define and redefine as the need arises. Moreover, the objects comprising the dialog manager eliminate the need for the application developer to manually define the dialog states and their interconnections.

These objects, along with the principles on which they are based, are the focus of this work. The semantics of the problem is not addressed in great detail but will be the focus of future work.

2 Natural Language Dialog Principles

In order to conduct a dialog there must be a reason to do so. We have defined seven such reasons. These reasons represent general principles behind a dialog. They are:

- Disambiguation of user inputs
- Relaxation

- Confirmation
- Completion of missing required information
- Description of otherwise invisible semantic actions
- Augmentation
- Detection of user confusion / Error Correction

The following are examples of dialog spawned by the above principles. Statements appearing in italic are user initiated statements and those in boldface are system generated statements.

Disambiguation

Disambiguation occurs if the system determines that something the user has said can be confused with something else that it knows about or if it can not execute the task in a unique manner.

Was Amadeus nominated for an academy award?
Is Amadeus the movie's title?
No.

The ambiguity in this example arises from the fact that Amadeus may be both the name of a movie and the name of a person therefore we must distinguish between the two.

Relaxation

Relaxation refers to the ability of the system to drop certain constraints that caused it to not be able to process the user's request.

I would like to travel first class from Newark to Miami tomorrow before 10AM.
There are no such flights. However, there are two flights that are not first class. Would you like me to list them?
Yes.
The departure times are 8:00 and 9:00.
This request was too restrictive and hence could not be fulfilled. The request must then be relaxed in order to find a flight. Offering a choice of business, coach or a flight after 10AM allows the collection of additional information that may make it possible to reserve a flight.

Confirmation

Confirmation is instantiated when a request could not be carried out. The system elects to confirm the value of a property to ensure that it understood it correctly. For example,

What is your id?
6473.
Upon not finding an id with the value 6473 the system will ask:

Is 6473 your id?

Completion

Completion refers to pieces of information that the dialog manager must know about in order to process the user's request.

I would like to travel from Newark to Miami tomorrow.
At what time?
In the morning.
First class, business, or coach?

In this example, the time of flight and the type of seating is required to make a reservation therefore it must be asked before searching for a flight.

Invisible semantic actions

The dialog system should alert the user of what we call invisible semantic actions. An invisible semantic action is an implicit action – one that may occur because of a constraint imposed by the user.

I would like a round trip ticket from Newark to San Francisco leaving tomorrow before 10AM.
First class, business, or coach?
Coach.
Returning when?
Next Friday after 6PM.
The only flight returning next Friday is a first class flight . Is this OK?

Rather than making the reservation knowing that the return flight is not coach even though the departure flight is ensures that the reservation conforms to the customer's request. Confirming whether or not the request is correct in this case is an example of an otherwise invisible semantic action.

Augmentation

The opposite of relaxation is augmentation. An under-constrained request will require augmenting the information that dialog system possesses in order to select among several possible solutions to the request.

Was Amadeus nominated for an academy award?
Is Amadeus the movie's title?
No.
There are several people with the name Amadeus. What is Amadeus' occupation?
Composer.

In this example there was not enough information to answer the initial question. The problem is that there is more than one person with the name Amadeus therefore more information about which Amadeus the user is referring to must be gathered.

User Confusion/Error Correction

If the user does not know how to answer a question the system must be able to ask another question in the hopes of executing the user's request. The obvious indication that the user is confused is if he/she answers a question with "I don't know" or something similar. The not so obvious indication is lack of progress, for example, if the system is having difficulties extracting a correct piece of information from the user.

Was Amadeus nominated for an academy award?
Is Amadeus the movie's title?
No.
There are several people with the name Amadeus. What is Amadeus'
occupation?
I don't know.
What is Amadeus' full name?
Wolfgang Amadeus Mozart.

This example resembles the previous one with the exception that when asked if the person knows Amadeus' occupation they do not. In this case another property of Amadeus must be found that may distinguish among the several people whose name is Amadeus.

Now that we have presented some dialog examples we will proceed with describing what it is that a dialog manager is responsible for.

3   The Dialog Manager

The dialog manager's ultimate goal is to manage a flawless conversation with a user in an attempt to satisfy the user's request. The user's request may be for a particular piece of information from a database (e.g. availability of a flight or car) or a command (e.g. to have a robot move an object to a desired location). Irrespective of the task the dialog manager needs to understand the user's request, know when to initiate a dialog, and choose what to say to process the user's request efficiently. Figure 1 is a pictorial representation of the dialog manager's flow of control. It begins when the user says something, either in response to an initial question posed by the system or as an initial request. The user's input comes into the system in the form of a sentence. The meaning is then extracted and placed in a semantic tree representation that the dialog manager manipulates and uses to determine if it needs to execute one of its dialog motivating algorithms. If it decides that it needs to ask the user a question it calls the natural language generation component. If the dialog manager determines that the request is not ambiguous and does not lack information it passes the tree representation along to the input filter.

The input filter's responsibility is to extract the needed information from the tree representation and put it in a format suitable for the application. The application then executes the request and sends back an appropriate response that is prepared by the output filter and sent back to the dialog manager. The response is sent back to the dialog manager to determine if the result of executing the user's request led to any ambiguities, in which case the process is repeated. If it does not find any ambiguities then the response is presented to the user.

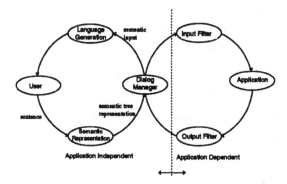

Fig. 1. When the user poses a request, the dialog manager must be able to detect whether the request can be carried out as is. If it finds it can not then it asks the user questions in the hopes of reaching a request that can then be passed to the application.

In posing a question to the user the system must select those questions that are *meaningful, efficient,* and *computationally inexpensive.* For a question to be meaningful it must be related to what the user wants. Efficiency means arriving at an appropriate request by asking the fewest number of questions. For a question to be computationally inexpensive means that it can easily be retrieved from the database if the request is for a piece of information, or that the task can easily be carried out if the request is a command. Determining if a question is computationally inexpensive is the application developer's responsibility.

The dialog manager consists of a set of objects for handling various key dialog principles. These objects are unaffected by the particular application. Application independence and scalability led to the development of the dialog manager's object-oriented and rule-based design. A finite-state approach requires that we define all possible dialog states and their interconnections. For large applications this may mean defining many states with complex interconnections. Redefining or augmenting such a network could be a formidable task. Our approach relies on the dialog manager's ability to determine which dialog principles to apply and when. This eliminates the need to predict all possible dialog states and their interconnections. This makes the application developer's job much easier.

Since the main emphasis of this paper are the dialog principles we will give only a brief summary of the underlying representations used by the dialog manager. They include a grammar, frames, and the interpretation tree.

Language Understanding. The first important step in conducting a good dialog is to understand what the other party has said. In order to enable the system to do this it is necessary to supply it with a grammar that will enable it to parse a variety of queries and requests. The grammar is written in a high-level language called Grammar Specification Language (GSL) [4]. In addition to specifying the syntax of the grammar we take advantage of the grammar compiler's capability of including semantics in the grammar. The semantics is key in building a proper interpretation of the user's request.

Frames. A frame is a collection of properties that describes an object or concept. For example, a person is an object, name is a property and John Smith is the value of a property. Frames may inherit properties from other frames. Each of the properties is assigned a weight. The weight conveys to the dialog manager the importance of a property. The higher the weight the more important the property. These weights are used by the dialog manager when it formulates its queries. It will decide to ask the user about that property which has the highest weight since this is an important property. An application builder must define those frames that represent the knowledge about the application. Figure 2 shows a hierarchy of frames that represents the concept of a reservation.

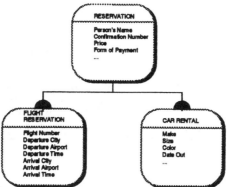

Fig. 2. Frames may inherit properties from other frames. In this example, the **flight reservation** and **car rental** frame inherit properties from the **reservation** frame.

The interpretation tree The grammar assists the dialog manager in building the tree structure that represents its interpretation of a sentence. In constructing the interpretation tree the dialog manager may find two or more possible meanings for a user's utterance. If this is the case it creates a tree with the same number of branches as possible interpretations for the sentence and attempts to disambiguate it. An example of an interpretation tree is given later on, in figure 3.

4 Disambiguation

There are essentially two situations that lead to disambiguation. The first occurs when the user has said something that is ambiguous and the second occurs when there are too many responses from the application. If either one of these situations should occur the dialog manager must be able to select a question to ask of the user in the hopes that an answer to that question will resolve the ambiguity.

4.1 Question Selection: Example

The question that the dialog manager selects to ask first can adversely affect the number of questions that it will have to ask. It is important to ask the fewest number of questions to avoid tedium. Consider the example of figure 1. Assume the dialog manager has detected an ambiguity and that the interpretation tree consists of five branches and three properties it can ask about. The frames represent the concept of a flight reservation and the three properties include the departure city (DepCity), the arrival city (ArrCity) and the departure time (DepTime). Table 1 illustrates the values of the properties for each of the five branches. We assume that the outcome of the selection process is equiprobable.

Table 1. This table depicts a scenario where there is an ambiguity in the interpretation tree and there are five branches contributing to the ambiguity.

	B1	B2	B3	B4	B5
DepCity	Boston	Boston	Wash.	NY	Wash.
ArrCity	Wash.	NY	NY	Wash.	NY
DepTime	9:00	9:00	11:30	11:30	9:00

If the dialog manager were to ask about the departure city first then the user may respond with Boston, Washington, or NY. To satisfy the equiprobability of outcomes, there is a 40% probability that the user will say Boston, a 40% probability that the user will say Washington and a 20% probability that the user will say NY.

If the user responds with Boston then the dialog manager must discriminate between the branches **B1** and **B2**. If the user responds with Washington then the dialog manager must distinguish between the branches **B3** and **B5**. However, if the user responds with NY then the dialog manager does not need to ask any further question because this narrows the search for the correct branch to one, namely branch **B4**. In essence asking a question reduces the number of branches contributing to the ambiguity.

If the user responds with Boston as the departure city then the dialog manager will choose to ask about the arrival city rather than the departure time because a response to the question about the arrival city will single out a branch, while a response to departure time will not because both flights have a departure time of 9:00. Likewise if the user responds with Washington as the departure city then the dialog manager will ask about the departure time because the arrival city is the same and hence it would not be able to distinguish between the two flights.

The expected number of questions if the dialog manager chooses to ask about the departure city first is 1.8. This value is computed based on the probabilities of the responses. In this situation the dialog manager must ask at least one question. After asking the first question there is a 40% probability that it will have to ask a second question should the user respond with Boston plus a 40% probability that it will have to ask a second question if the user responds with Washington plus 0 if the user responds with NY since this selects one branch. Therefore the total expected number of questions is 1.8 $(1 + .4 + .4)$.

Had the dialog manager chosen to ask about the arrival city first instead of the departure city the total expected number of questions would be 2.4. The reason for this is that there is the high probability that the user will respond with NY to the question of arrival city. If this is the case then the dialog manager may have to ask two more questions. If the user responds with NY when asked for the departure city then there is a 67% probability that the user will respond with Washington which would mean asking yet another question, namely the departure time before being able to distinguish among all five branches $(1 + .4 + .6 * (1 + .67)) = 2.4$.

4.2   Computing the minimum expected number of questions

We have seen how the initial query property can affect the number of subsequent queries. The dialog manager determines the initial query property by computing the expected number of subsequent queries for each property. Let us define the following:

- B: set of branches
- p: a frame property
- P: set of p's
- w: weight associated with a property, $w \geq 1$
- V: set of different values of property p in B
- S: set of instances that have a given value for a given property p
- $\frac{\|S(v,p,B)\|}{\|B\|}$: probability that instance B has value v for property p.

The dialog manager uses the following recursive criterion to compute the minimum expected number of questions:

$$N(B, P) = \min_{p \in P} \left\{ \frac{1}{w} + \sum_{v \in V(p,B)} \frac{\| S(v, p, B) \|}{\| B \|} N(S(v, p, B), P \backslash \{p\}) \right\} \quad (1)$$

It chooses a property $p \in P$ as the first property and recursively calls itself on the remaining properties computing how many expected questions remain to be asked. It computes this number for all properties $p \in P$ and chooses the p that yields the minimum expected number of questions.

4.3   Question Selection: Property Categorization

Recall that properties are assigned a weight to measure their importance; the larger the weight the more important the property. In addition, properties are also assigned a state; they are either static or dynamic. Static properties are those properties that are defined as part of the application definition. They are either approximate properties like a person's name (the user may not refer to the person's entire name but rather only to their first name, which is what would make this property approximate) or properties that are computationally inexpensive to retrieve from the application. We will call the approximate properties S_1 and the inexpensive properties S_2 and all other properties S.

The dynamic properties are those that are determined based on the current context of the dialog. These include those properties bound in the context of the request (D_1), properties requested by the user (D_2), and all other properties that are not static (D).

$S_1, S_2, D_1, D_2, D,$ and S are all sets of properties. These sets, combined and intersected, form four distinct categories that provide a useful measure of the overall importance of a property in the current context. Properties that fall into the first category C_1 are the most attractive properties to use when the system needs to formulate a query. It is defined as the set $(S \backslash S_1) \cap S_2 \cap D$. This category consists of those properties that are exact, inexpensive, and not bound. This set is the most attractive because exact properties are more likely to be stated without ambiguity. An inexpensive property has the obvious advantage of being easy to retrieve from the application, asking about an unbound property ensures that the question formulated is intuitive.

The second category C_2 is defined as consisting of those properties that are approximate and bound, $S_1 \cap D_1$. Even though this set consists of properties that are bound it also refers to those that are approximate and hence may require clarification.

The third category, C_3, is defined as $S_1 \cap S_2 \cap D$ which consists of those properties that are approximate, inexpensive and not bound. This set gets third priority primarily because it refers to properties that are not bound and approximate which have a higher probability of leading to ambiguities.

Finally the fourth category C_4 is the least attractive, it is defined as $S \backslash (C_1 \cup C_2 \cup C_3)$ which means all other properties. Table 2 summarizes this information.

For example, the arrival city is an approximate property of the Flight class, while departure time is an exact property. It is more prudent to ask about the departure time first, since we believe that this information will lead to disambiguation with more certainty.

In our criterion we added $\frac{1}{w}$ to the sum to account for the initial question and the weight associated with the property being asked. The weight associated with a property is a global measure of its importance but it does not capture its local importance. A property's local importance is its importance in the current context of the dialog. Membership in $C_1, C_2,$ and C_3 provides a way of combining the local and global importance of a property. To do this we define the weight

Table 2. A summary of property categorization.

		exact, (not approximate), inexpensive, not bound
C_1	$(S \backslash S_1) \cap S_2 \cap D$	exact, (not approximate), inexpensive, not bound
C_2	$S_1 \cap D_1$	approximate, bound
C_3	$S_1 \cap S_2 \cap D$	approximate, inexpensive, not bound
C_4	$S \backslash (C_1 \cup C_2 \cup C_3)$	all other

in our criterion to be $w = w_p + w_c$, were w_p is the weight associated with a property and w_c is the weight associated with a category. For example, category C_1 has a higher weight than C_2 and C_3.

4.4 Collapsing

We have seen that when faced with an ambiguity the dialog manager must solicit pieces of information from the user in its attempts to clarify an ambiguity. It accomplishes this by asking the user questions. What follows is the utilization of the user's response to these questions to clarify an ambiguity. In other words, selecting a branch from the interpretation tree. The process by which the dialog manager does this we call *collapsing*. When the dialog manager presents the user with a query it creates an *expectation tree*. The expectation tree is the depository for the user's response.

The expectation tree like the interpretation tree consists of nodes that are instances of frames. When the dialog manager has decided what property to ask about it forms an instance of the appropriate frame. For example, if the dialog manager asks **What city does the flight depart from?** it creates a flight instance and marks the departure city as the property for which we expect a response from the user.

Collapsing: Example Let us assume that the user has asked to reserve a flight from NY to Boston departing in the morning and that the application has returned three such flights leaving in the morning. The interpretation tree is depicted in figure 3. The question selection policy returns the departure time as the property it should ask the user to clarify since this property differs among the three flights and it is a property which is important from the standpoint of reserving a flight, certainly more so than the aircraft. The system responds with: **There are 3 flights. The departure times are 7:00, 8:00, and 9:00. Please select one.** At this point the system creates the expectation tree which consists of a flight instance whose departure time is marked as being the property we expect the user to fill in as a result of the system's question. Assume that the user selects 8:00 as the departure time. The system will proceed to collapse the expectation tree and the ambiguous tree that contains the three flights. It performs a pattern match among the properties of the expectation tree and the

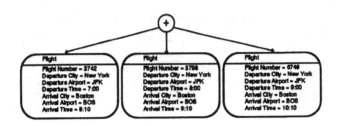

Fig. 3. The ambiguity arises from the fact that there are three flights that satisfy the user's request.

ambiguous tree.[4] It will select the branch of the tree whose departure time is 8:00 since this matches the departure time in the expectation tree.

If the user says something that the dialog manager is not expecting in its expectation tree then it will confirm the value the user said. If what the user said is indeed what the system said then it will proceed to ask another question in the hopes of clarifying the ambiguity. If it runs out of questions it notifies the user that it can not carry out the request.

5 Relaxation and Confirmation

There will be occasions when, even with an unambiguous tree and no missing information, the dialog manager may not be able to fulfill a user's request without further intervention. We have seen what the dialog manager does when it encounters an ambiguity in the user's request. What we have not seen is what it does when it is faced with a request it can not execute. When this occurs the dialog manager does one of two things, it either confirms or relaxes. To confirm means to verify the value of certain properties associated with the request. To relax means to drop certain properties associated with the request.

Before relaxing a constraint the dialog manager may choose to confirm the value of a particular property. It first chooses to confirm those properties that are approximate. These properties have a higher probability of being misspoken or misrecognized. If confirming a property results in the realization that the value is correct then the dialog manager proceeds to relax a constraint. If confirming a property results in the realization that the value is incorrect then the user is prompted for the value of that property again, and the dialog process is repeated.

The issue with relaxing constraints is knowing which constraints to relax. It will certainly not drop any required properties since these are vital to fulfilling the user's request. It will begin by dropping those constraints that have low weights. If this results in retrieving a tuple then the information is presented to the user. If the system runs out of properties it can drop then it will choose to confirm those properties that have high weights.

---

[4] The expectation tree is an implementation necessity; it is a placeholder for a user's response in a form of a tree.

6   Completion

We have now seen how the dialog manager disambiguates, relaxes and confirms. The remaining functionality is *completion*. Completion falls into the category of traditional question-answer systems. Completion is simply querying the user for any missing required information. The dialog manager determines that there is missing required information by examining the weights associated with the properties of those frames that have been instantiated. If there are properties lacking a value and whose weights are sufficiently high then the dialog manager will query the user for the values.

7   Case Study: A flight reservation system

We will illustrate the dialog system's managing skills on a flight reservation system. The goal of the flight reservation system is to reserve flights for airline employees. In this application the dialog manager is supplied with a scenario. The scenario provides the dialog manager with the information it needs to understand and execute the user's request. The statements in italic are the user's and those in boldface are the system's. It begins as follows:

Welcome to the flight reservation system. What is your id? *456.*
If the id is not found the system will confirm the id.

Is 456 your id? *Yes.*
Since the system understood the correct id it will try to ask the user another question namely,
What is your last name? *Brown.*
Again the system seeks information from the database of an employee whose last name is Brown. In this case it retrieves two people whose last name is Brown. This causes the dialog manager to apply the disambiguation principle and the system asks:
What is your first name? *Mike.*
Had there only been one Brown then it would not have asked about the first name. But since there was more than one person with the name Brown it needed to clarify which one. It didn't ask about the id number again because it had already confirmed it, but could not find it in the database. The only property left to ask about was the caller's first name. This finally discriminated between the two people whose last name was Brown. The system did not ask for the user's first name, which would have meant asking one fewer question because the weight associated with the caller's last name was higher implying more importance.

Had the user not known their id to begin with, for example,

What is your id? *I don't know.*
Then the system would have asked

What is your last name?
Once it has verified the caller's identification it proceeds to examine the script for further instructions and asks

What is your departure city? *New York.*
What is your expected time of departure? *In the morning.*
What is your destination? *Boston.*
Once it has this information it submits the request to the database. For this particular example the application returns two flights departing from New York in the morning and arriving in Boston. Therefore it generates the following statement

There are two such flights. The departure times are 8:00 and 9:00. Please choose one. *8:00*
The dialog system will list up to three possible choices to choose from. Should there be more it would say

There are many flights. Please more accurately specify the departure time.

The dialog manager chooses to ask about the departure time because it differs among the two flights and because it has a high weight meaning that it is an important property and likely to aid in the disambiguation process.

This dialog scenario defined a form-filling dialog where the system has all the initiative and asks the user a series of questions to solicit a single piece of information. The scenario can also specify a dialog that is more flexible and gives the user more of an initiative. Work on how the scenario is defined and what information it requires is ongoing.

8 Concluding Remarks

Building a working dialog system capable of conducting a conversation we consider natural is certainly a challenge. It embodies many of the important problems facing natural language processing. We showed in this paper our efforts in constructing such a working dialog system. We began by defining those dialog principles we feel are important and designed a set of objects based on these principles. Part of building a working system meant constructing a working dialog system utilizable on a wide range of task semantics. This was accomplished by choosing frames as the representation of the task. The dialog manager then uses these task definitions to initiate a dialog motivated by the dialog principles.

The work presented in this paper does not claim to have developed a model that integrates all the necessary constituents of natural language processing into a system capable of discriminating every ambiguous, ill-formed, incomplete sentence that we as humans make in our everyday conversation. What it does present is a system that can conduct an intelligent dialog under the guidance of the application developer. The system, in its prototype form, is currently being integrated into various automated telephone services.

References

1. Harald Aust, Martin Oerder, Frank Seide, and Volker Steinbiss, 'The Philips automatic train timetable information system', *Speech Communication*, 249–262, (1995).
2. D.G. Bobrow, 'GUS: A Frame Driven Dialog System', *Artificial Intelligence*, 155–173, (1977).
3. L. Bolc, A. Kowalski, M. Kozlowski, and T. Strzalkowski, 'A natural language information retrieval system with with extensions towards fuzzy reasoning', in *International Journal Man -Machine Studies*, (Oct. 1985).
4. M.K. Brown and B.M. Buntschuh, 'A new grammar compiler for connected speech recognition', in *ICSLP94*, (Sept. 1994).
5. Sandra Carberry, *Plan Recognition in Natural Language Dialogue*, The MIT Press, 1990.
6. Robin Cohen, Ken Schmidt, and Peter van Beek, 'A framework for soliciting clarification from users during plan recognition', in *Fourth International Conference on User Modeling*, (1994).
7. Robert E. Frederking, *Integrated natural language dialogue: a computational model*, Kluwer Academic Publishers, 1988.
8. B.J. Grosz, D.E. Appelt, P.A. Martin, and F.C.N. Pereira, 'TEAM: An experiment in the design of transportable natural language interfaces', *Artificial Intelligence*, **32**, 173–243, (1987).
9. G.G. Hendrix, E.D. Sacerdoti, D. Sagalowicz, and J. Slocum, 'Developing a natural language interface to complex data', *ACM Transactions on Database Systems*, 105–147, (June 1978).
10. Bhavani Raskutti and Ingrid Zukerman, 'Query and response generation during information-seeking interactions', in *Fourth International Conference on User Modeling*, (1994).
11. S. Seneff, 'TINA: A natural language system for spoken language applications', *Computational Linguistics*, 61–86, (1992).
12. Ronnie W. Smith and D. Richard Hipp, *Spoken Natural Language Dialog Systems: A Practical Approach*, Oxford University Press, 1994.
13. R. Wilensky, 'The Berkeley UNIX Consultant Project', *Computational Linguistics*, **14**, 35–84, (1988).
14. S.J. Young and C.E. Proctor, 'The design and implementation of dialogue control in voice operated database inquiry systems', *Computer Speech and Language*, 329–353, (1989).
15. S.R. Young, A.G. Hauptmann, W.H. Ward, E.T. Smith, and P. Werner, 'High level knowledge sources in usable speech recognition systems', *Communications of ACM*, 183–194, (February 1989).

Designing a Portable Spoken Dialogue System*

James Barnett[1] and Mona Singh[1,2]

[1] Dragon Systems Inc.
320 Nevada Street
Newton, MA 02160, USA
[2] Department of Computer Science
North Carolina State University
Raleigh, NC 27695-8206, USA

Abstract. Spoken dialogue systems enable the construction of complex applications involving extended, meaningful interactions with users. Building an effective, generic dialogue system requires techniques and expertise from a number of areas such as natural language, computer-human interaction, and information systems. A key challenge is to design a system through which user-friendly applications can be created in a developer-friendly manner. We present the architecture of Dragon Dialogue, which is a domain and application independent spoken language system. The architecture uses limited knowledge, encapsulated in separate domain, transaction, and dialogue models. These models help customize this architecture to new applications, and facilitate limited but robust natural language techniques for natural prompt generation. We adapt ideas from information systems to handle recovery from errors and commitment in a principled manner. Consequently, our approach marries user-friendliness with developer-friendliness and incorporates sophisticated functionality without increasing the demands on the application developer.

1 Introduction

As speech recognition systems are becoming more common, attention is shifting to speech dialogue systems. These are systems that enable the construction of complex applications involving prolonged meaningful interactions with users. Dialogue systems come with their own set of research and implementational challenges beyond speech recognition. We are developing a system in which the dialogue strategy is domain-independent for the most part, and which can be set up by an application developer for different applications. Thus the architecture proposed in this document aims at maximizing the reusability of the dialogue system. The developer provides the various models through which our architecture is instantiated. Our challenge is to define a formal specification and technique through which each of these models could be defined and used. We have selected

---

* We would like to thank Dr. Paul Bamberg, Professor Munindar Singh, and the anonymous referees for their comments on this paper.

information retrieval and structured database access as our initial testbed applications because they provide a rich structure and yet are tractable [Singh & Barnett, 1996].

Current research systems are limited in their natural language and dialogue management capabilities, because their prompts are either fixed or chosen from a small predetermined set [Waterworth, 1991; Proctor & Young, 1991]. We propose robust natural language techniques based on limited domain models, which enable us to improve naturalness without increasing our demands from the application developer. The relevant dialogue components include prompts and recognition and generation grammars. Automatically generated prompts can be based on the current state of affairs and history of a particular dialogue. Thus, they can be more to-the-point and less repetitious.

Another crucial issue is the amount of domain-specific knowledge that must be encoded in the application. Some spoken language systems utilize a deep domain model. Such elaborate hand-crafted models are developer-unfriendly in that they are hard to construct and maintain. Most other systems have almost no model of the domain. These are easier to develop, but prove rigid in usage. We choose a trade-off, i.e., limited models, that are not hard to build or maintain but are still highly beneficial. Given the state of the art in knowledge representation, anything beyond limited models would not be practical for a system that is intended for wide use in varied domains. Therefore, our design focuses on the semiautomatic generation of speech interfaces from restricted amounts of domain knowledge. We use the levels of domain knowledge typically required by systems for accessing information from databases and open environments. This level of domain knowledge is essential for semantically modeling information sources in order to access them effectively.

The rest of the paper is organized as follows. Section 2 presents a brief introduction to the architecture of Dragon Dialogue. Section 3 outlines the dialogue control strategy of our system. An example dialogue is given in Section 4. Section 5 presents our conclusions and suggests future directions that would improve the quality and usability of spoken language dialogue systems.

2 The Architecture

The dialogue system described in this paper is a *mixed-initiative* system—i.e., one in which both the user and the system can exercise control. The input and the output modality is limited to speech in the present version. However, we envision incorporating other modalities (e.g., keypads and graphical displays), as the system becomes more stable. Our system is declaratively structured through separate models for each major module. We cover the first three in greater detail in the present paper.

- *Domain model:* Provides a static model of the domain. This describes the class of representations that are built up through each dialogue. This model is also used to generate natural language prompts, clarificational dialogue, and recognition grammars automatically.

- *Transaction model:* Provides the possible domain-specific interactions. This shows how the steps are interrelated, and represents the key properties of the backend system on which the transactions are ultimately effected.
- *Dialogue model:* Provides the linguistic realizations of the transaction model. It matches the incoming strings to the slots, decides the conversational move to be made and produces the appropriate linguistic string.
- *User model:* Provides the expertise of the user in engaging with the system.
- *Speech recognition model:* Provides the level of confidence in the recognized speech input.

2.1 Domain Model

The domain model refers to a declarative description of objects and relations in a particular domain. We use the standard Entity-Relationship (ER) approach, augmented with action descriptions, for specifying the conceptual model for a domain [Batini *et al.*, 1992]. The ER approach diagrammatically captures a high-level understanding of the data independent of its physical storage in the database. Entity-relationship diagrams are widely used in the information systems community. Database designers are familiar with them and often databases have a conceptual view represented as an ER diagram. In addition, since the ER approach has become quite standard, there are various commercial products available that allow the developer to graphically create an ER model for a domain.

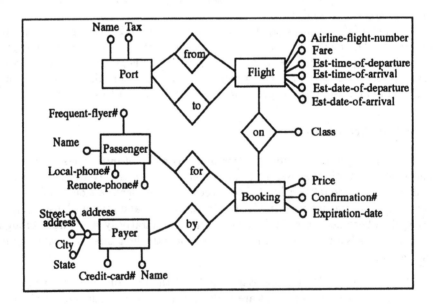

Fig. 1. Example Domain Model

The ER approach has many advantages. A conceptual model of the data is easy to understand and can be used to communicate with non-technical users. The basic concepts of ER diagrams include *entities, attributes,* and *relationships.*

- Entities are the basic concepts of an ER model. They represent real and abstract objects that are pertinent. Examples of entities in Figure1 are *passenger, booking, payer, flight,* and *port.*
- Attributes are properties of entities. For example in Figure 1 *payer* is an entity that may be described by attributes such as *name, address,* and *credit card number.* Attributes can be composite. The composite attribute *address* has be subdivided into the *street-address, city, state,* and *zip-code.*
- Relationships relate entities in the conceptual schema. Relationships may vary in the number of participating entities. For example *for* is a binary relationship. Relationships, e.g., *on,* like entities, may have attributes.

We also incorporate subclasses and superclasses. For example, instances of the passenger entity type may be grouped into *first-class passenger, business-class passenger,* and *economy-class passenger.* Every instance of the entity *first-class passenger* is also a member of the entity *passenger.* The *passenger* entity type is called the superclass and the *first-class passenger, business-class passenger,* and *economy-class passenger* entity types form the subclasses. An entity in the subclass inherits all the attributes of the entity in the superclass. The subclass entity also inherits all relationship instances in which the superclass participates. Incorporation of these constructs facilitates the natural language component.

2.2 Transaction Model

The transaction model refers to a model of all domain specific interactions between the system and the user for a particular application. The transaction model includes a template that the application developer designs. The template is like a form that needs to be filled in during the dialogue. It encodes all the information required for each transaction type in that particular application. Consider a form with a slot for *name.* Though this slot is quite basic and common, a certain format is required for the subslots. The form may specify a place for the *last name* and one for *first name.* This specification may vary from one domain to another. In the case of an application for immigration or security clearance there may be a subslot for *all previous names, maiden name,* and *aliases.* The user must specify the input in that particular format. In addition, the transaction model keeps track of the slots that are absolutely vital for the form to be complete. Thus while the slot *maiden name* is not vital for an airline reservation, it is vital for immigration.

The transaction model keeps track of the relationship between the type of transactions and the slots required to complete the transaction. In the airline reservation example presented in Figure 2, four slots, namely, *payer name, credit card number*, *expiration date*, and *payer address* are vital for buying a ticket while *frequent flyer number* and *local phone number* are not.

Via the template the developer specifies the information that is required for the different transactions and the dialogue that takes place between the system and the user is geared towards getting all the crucial information in the most efficient and user-friendly manner.

The transaction model deals with the specified domain model and the parts of it that are desired for a particular transaction. This enables various efficiencies to be implemented in the system. For example, the system allows slots to be specified in any order by the user and allows the interactions to proceed in a user-controlled or a system-controlled manner. It is up to the developer to set the slots in any way he likes. An application where the slots are set in a strict order would yield a system-centric dialogue application. In the case of user-centric and mixed-initiative systems, subtemplates may be ordered while the order of the slots within them may be free. For example, in making an airline reservation, the subtemplates for passenger information must be ordered after the *from port* and *to port* slots. However, the slots *from port* and *to port* need not be mutually ordered. The transaction model in this case is more like the set of weakest allowed ordering constraints.

Figure 2 diagrams a transaction model for a greatly simplified version of air travel transactions. This application includes several transactions such as *getting a price quote*, *booking a ticket* and *buying a ticket*. The transactions that a user cares about are mapped into transactions of the backend system.

- *getting a price quote:* vital information includes the departure time, date and port, arrival time, date, and port. The returned value is the fare.
- *booking a ticket:* vital information includes the departure time and port, arrival time and port, date of travel, name of passenger. Non-vital information includes the address of the passenger, phone numbers in departure city and in the arrival city. The returned value at the end of a successful transaction is the confirmation number and a price quote.
- *buying a ticket:* vital information includes the departure time and port, arrival time and port, dates of travel, name of passenger, address of the passenger, credit card number. Non-vital information includes the phone numbers in the departure city and in the arrival city. The returned value at the end of a successful transaction is the confirmation number.

Our approach also incorporates transactional notions of commit, rollback, and checkpoint from databases. The main question that arises while interacting with information systems is knowing how actions by the user are committed to the system. For example, if you attempt to buy an airline ticket, how does the system know you really decided to buy it? More mundanely, many current voice-mail systems allow people to call in and then selectively review, delete, and forward messages. These systems are not speech based, but the point carries over. When is a message really forwarded? Similarly, when a message is deleted it may either evaporate immediately, or the system might give the user an opportunity

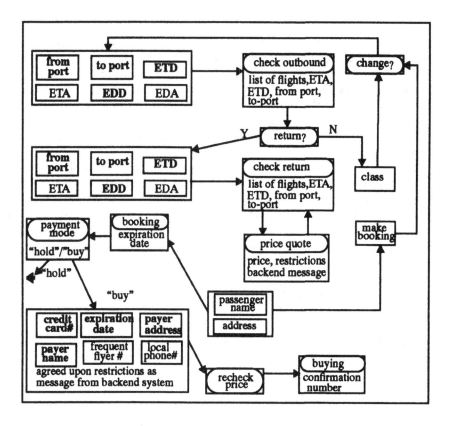

Fig. 2. Example Transaction Model

to decide again whether to expunge it. Unfortunately, many voice mail systems treat a disconnect, even an involuntarily disconnect, as equivalent to a commit! This is the sort of behavior that makes interfaces unfriendly.

In dealing with information systems checkpoints are crucial for error recovery in complex transactions. If the user has engaged in a long interaction with the dialogue system and makes or detects an error, how much of his interaction can he salvage? The user would like to be able to selectively undo certain changes or inputs. However, the information system may not allow arbitrary undo actions. A practical compromise is that certain appropriate states for checkpoints are defined by the application programmer. The user has the option of setting checkpoints after a certain semantically appropriate subtransaction has concluded. If the next subtransaction runs into trouble, the user can ask the system to rollback the most recent changes and go to the previous state that was valid. In this way, errors cause only a small amount of work to be lost.

We believe that an effective dialogue system will have the abstractions for commit, decommit, checkpointing, and rollback. These abstractions will cover both the backend system aspects and the user issues, meaning how users know what is available, how they are constrained, and what their choices are. The dialogue system will provide these abstractions to the system programmer to decide where and how to incorporate them. Only the programmer can decide what should be done in a specific application and what view should be presented to what class of users. Once the required information has been obtained from the user, i.e., all the vital slots in the transaction model have been filled, an appropriate request is generated and passed on to the backend system. Appropriate dialogue is generated by the dialogue model to make the commit points known to the user.

The developer thus creates nodes and slots in the form of templates. Each node is initialized by its name-value pairs. Slots are a special case of nodes. As the dialogue progresses other information is added to each node. The confirmation/verification status of the node is important for various reasons. First, it provides flexibility to the system in carrying out confirmation and verification depending on, say, the level of confidence in the recognition of the user input. The transaction model may include a default set of rules. For example, commit nodes must always be confirmed. However, a system developer could set it up so that in cases when the confidence level for the input of a vital slot is low the slot value is verified right away rather than waiting for confirmation at a higher level. Second, in complex transaction models, the same slot may be used more than once. Not having the user verify the same thing over and over would certainly be user-friendly. Thus we keep flags on slots regarding their status.

2.3 Speech Recognition Model

A measure of the level of confidence in the speech recognized is a hard but useful research problem today. The basic intuition here is that the system does not always perfectly understand what the user said, so the dialogue model's actions may be chosen to minimize the damage this can cause. We believe that if the speech recognition system returns the level of confidence along with the user's input, the confidence level could be used in several ways to optimize the dialogue strategy. Suppose recognition could be characterized as being good, poor, or somewhere in between. If the speech input is recognized with a high level of confidence, the confirmation of the slot may be put off till the current part of the dialogue is complete; on the other hand, if the recognition of speaker input is poor it may be appropriate to verify the input before proceeding further.

There are several instances where we believe that the confidence in speech recognition is important. We present a few examples here. In section 3 we give an example rule that is based on the confidence level of the recognized string.

- In the case of poor recognition, it may be possible to fill in a slot based on another slot using constraints among the slots. For example, in the airline reservation domain, the name of payer could be the same as the *name of passenger* for whom the reservation is being made. The filled-in default value may be confirmed later.

– The developer may specify the shortest routes, or alternative paths that could help complete the transaction in case of poor recognition. Thus, it is possible to control the feedback frequency to make the dialogue more natural.

– If the recognition is poor, the user may just be switched to a human operator. This may seem like a trivial example but is extremely crucial in the usability of the system. If the user is having difficulty in a successful dialogue because the user is a novice then we need to provide appropriate feedback as we switch the user to a human operator. If the recognition is bad then we could provide feedback that would help the recognition. Thus feedback could include asking the user to speak closer or farther from the microphone and so on. In such cases we could switch to the telephone keypad or the computer keyboard instead of switching to the operator.

These cases are examples of the ways in which we think a speech recognition model may help the dialogue. They are merely examples of how we think the confidence in speech recognition can be used for a user-friendly interface.

2.4 Dialogue Model

The dialogue model has several important functions. First, it attempts to find a match between the slots in the templates for the transaction types based on user input. For example, if a user calls in and says "I want to go to Chicago today," the role of the dialogue model is to pick out the slots and their values. Second, some slots may be marked by the developer as requiring clarification. In such cases the dialogue model generates appropriate clarifications based on the information provided by the user.

Reichman proposed a set of conversational moves [Reichman, 1986]. Based on the intuition that the conversational moves that take place in a task-oriented setting, especially in highly specialized spoken language interfaces, are very different from the conversational moves of informal communication we propose the following conversational moves for task-oriented spoken dialogues: *request for information, information, request for verification, verification, request for confirmation, confirmation, paraphrasing, agreement, restatement, further development, return to previous context space, interruption, feedback, presentation of data, correction, explanation,* and *silence.* We distinguish between three kinds of clarificational conversational moves *information, verification, confirmation,* and *paraphrasing.*

– Information refers to the basic request for information that is required for each slot (and subslots). These prompts may be provided by the developer and are used to request information about a particular slot from the user. The developer may provide multiple prompts for a slot. For example, for the destination slot the developer may provide the prompts "What is your destination?" and "Where would you like to go?" and so on.

– Verification of speaker input is instantaneous and direct. For example, if the user says "I want to go to Boston" the verificational dialogue would be echoed back immediately and could be of the form "Please repeat the name of your destination city."

- Paraphrasing, like verification, is instantaneous but the value of the slot is a part of the system prompt, e.g., "Is your destination city Boston?" The choice of the verificational dialogue depends on a number of factors such as the level of confidence in the speech input and the confusability of the subslots [Proctor & Young, 1991]. Slots that are considered highly confusable by the developer may be set up for paraphrasing instead of verification since it is likely to resolve ambiguity more reliably [Tanaka *et al.*, 1996].

- Confirmation refers to indirect and delayed clarification. Confirmations are rolled into whatever is the current response. For example, when the system returns the price quote to the user, it could confirm the vital slots by repeating the values of those slots, e.g., by saying: "The fare for travel from Boston to Chicago leaving on the eleventh of March is two hundred and seventy dollars."

The conversational moves of correction, restatement, and further development are standard. Correction provides a new value for a slot. Restatement and agreement do not change the existing value of a slot, and further development results in the standard procedure of going to the next set of slots. Feedback refers to the system relaying its status to keep the user informed of delays. Feedback is believed to lower user frustration [Shriber *et al.*, 1992]. The conversational moves of silence and interruption present more interesting challenges to a dialogue system. Silence refers to the lack of input from the user after he is prompted for a response. Interruption is more complicated. Since spoken language dialogue systems are task-oriented, interruption in these systems may be very different from interruption in human-human conversation. Interruption in our system is classified according to when it occurs. A user may interrupt while being prompted for information, verification, or confirmation. Another kind of interruption may occur when the results are being presented to the user. Many speech recognition systems don't allow barge-in, so the conversational move of interruption is not a possibility in those cases.

The run-time dialogue model consists of dialogue nodes (Dnodes) linked to appropriate transaction nodes (Tnodes) and action nodes (Anodes). The Dnodes correspond to the different conversational moves that are possible in the dialogue. Dnodes are automatically generated for each node in the transaction model. A Dnode contains the dialogue components and has its grammar built from the grammar of the corresponding nodes or simple keywords.

3 Dialogue Control Strategy

As explained above, various models are specified formally. We have made the system modular and declarative. The basic control loop (outlined in Figure 3) that interprets the representations is quite simple. After initialization, the system alternates between a *user phase* and a *system phase*.

```
main ()
{configure();
  repeat
    if user-allowed? then stop = userPhase();
    if system-allowed? then stop = systemPhase();
  until stop
  cleanup();
}
```

Fig. 3. The Main Loop

The user phase (given in Figure 4), receives an utterance from the user— it invokes the recognizer with a list of possible grammars for the nodes in the transaction model. The recognizer produces a list of interpretations along with the matching grammar and the confidence in the match. The system picks out the node yielding the best interpretation. The best interpretation is chosen by a metric that prefers the best confidence and highest probability of occurrence— the latter is based on the probability of a particular slot being the next slot. From the selected slot, the system applies the rules describing the user interactions (these are coded in the user-rules ruleset).

```
userPhase (history)
{while (user speaks)
    u = get_utterance ();
/* u=(<u1, c1, g1>, ...,<un, cn, gn>)  here u is the utterance, c is
the confidence level of the speech recognized, and g is the grammar */
    for each ui
       node = {n:(curr_prob[n]* n->grammars_match_prob (ui)) is max}
       //the most likely-cum-best-matched node for the utterance
       node->apply_strat (history, user_rules)
}
```

Fig. 4. The User Phase

After a starting prompt for initialization, the system enters the "system phase." In this phase, the system selects the best target action node based on the current probabilities. The Anode determines the action chosen by the user. Notice that although in the user phase an Anode or a slot may be selected, in the system phase only an Anode may be selected. For example if a user says he is going to Boston on May 6 and from that we may figure he wants a price quote, but we prompt him based on what we believe he wants done. Once the target Anode is determined, the system checks if enough is known to execute the corresponding action. Specifically, if the vital inputs for the action are not known (at a high enough confidence level), the system attempts to obtain those values from the user. When the values are obtained, the system executes the appropriate procedure and presents the results. The execution can involve checkpointing and committing, which as we explained previously gives the user an opportunity to cause backend actions in a controlled manner. The presentation of results is also non-trivial, especially when the results are long or structured—we shall not explore the issue here. The system phase is briefly outlined in Figure 5 below.

```
systemPhase(history)
{update_target_probs()
/* take unfilled slots and their probabilities as the new target
probabilities */
  a = select_target(history, system_rules)        //an action node
  a->apply_strat();
  for each v belonging to a.vitals
    if v is not filled in
      then v->apply_strat(history, system_rules);
    a-> execute (history);
}
get_utterance()
{ recog = SDAPI_UTTERANCE ();  }
execute()
{apply_checkpoint();
  apply_commit();
  result = invoke_backend_proc();
  present_to_user(history, result);
}
```

Fig. 5. The System Phase

The apply-strategy module takes a node, a history, and a rule-set to determine what to do. The rule-set declaratively captures the intended behavior of the appropriate component of the system. The rule-set is matched against the history to find the best (most specific) match. The consequent action of this rule is executed on the given node. We allow actions to verify, confirm, commit, checkpoint, etc., which may involve a Tnode as well as a Dnode.

History matching is an interesting research issue. We allow rules with antecedents that have a predicate-variable structure with limited support for temporal aspects. The details of matching are beyond the scope of this paper but we present a partial set of rules for verification, simplified and sanitized for ease of exposition. In our history matching algorithm the most specific rule fires when more than one potential rules are in question.

– If you are at a node and the confidence level is <.5 and the node is verifiable,

```
apply_strat(history, rule_set)
{rule = most_specific_rule (history);
  if rule.action
    op = rule.action.operation      // every action has an op
    Tnode = rule.action.node        // every action has a Tnode
    script = rule.action.script
    case op
      VERIFY: if script == NULL then script = node.vscript;
                    verify(node, script)
      CONFIRM: ...
}
```

Fig. 6. The Apply Strategy Module

then verify the node. This rule can match any node whose name is bound to ?node. ?value is its likeliest value and ?conf is the corresponding confidence in the speech understood. This is the most basic rule that fires for all verifiable Tnodes.

```
if (Tnode ?node (?value ?conf))
    AND (?conf < 0.5)
    AND (verifiable?  ?node)
then VERIFY ?node
```

– If you verify a node and the response is "NO" with >.8 confidence level or the confidence level is <.2 then verify the node again. This rule can match any Dnode of the type VERIFY whose corresponding Tnode is bound to ?node, likeliest value is bound to ?yn and its confidence level to ?conf. This rule is more specific in nature than the one given above.

```
if (VERIFY ?node (?yn ?conf))
    AND ((?conf >0.8 AND ?yn=''NO'')
        OR (?conf <0.2))
then VERIFY ?node
```

– If you verify a node twice and the confidence level of the response "NO" the second time is >.8 or the confidence level in the recognition of the second response is >.2 then exit the node. Here exit refers to being forwarded to a human operator. This node can match any node such that the present and the previous nodes in the history of the system are both Dnodes for the given node. ?yn1 and ?yn2 are bound as before, as are ?conf1 and ?conf2.

```
if (seq (VERIFY ?node (?yn1 ?conf1))
        (VERIFY ?node (yn2 ?conf2)))
   AND ((?conf2 > 0.8 AND ?yn2=''NO'')
   OR (?conf2 < 0.2))
   then EXIT ?node
```

Thus rules that match the dialogue history get executed and determine what the next conversational move should be. The application developer may modify the set of rules and state new ones given the rule language. Thus the behavior of the system may be controlled very intricately by the developer. We want the developer to have such control because different applications require different system behavior. Representing conversational moves as a set of rules allows better portability of the system.

4 An Example

We now present a sample dialogue that illustrates how our system would work. Comments are included as indented plain text.

> User calls in.

Dragon Dialogue: *Thanks for calling the airline reservation service. You may book airline tickets within the United States. What are your ports of departure and arrival.*
User: *I would like to go to Boston on the 13th.*

> The system has the probabilities for the most likely slots that may be filled in at any point. The grammar of the user utterance matches that of the to-port slot and the estimated date of departure slot.

Dragon Dialogue (option 1): *Please repeat the arrival city?*

> The slot for the departure city requires verification. However, if there is a rule like the one presented above whereby verificational prompts are upgraded to confirmational prompts in cases where recognition is good. In that case, we would get the following alternative prompt.

Dragon Dialogue (option 2): *Where are you flying to Boston from?*

> Confirmation rolled into the prompt for the next slot.

User: *From Austin.*

> Dragon Dialogue checks with the backend system to see if there are any flights from Austin to Boston. The response is positive so the dialogue continues.

Dragon Dialogue: *What is the date of departure?*

Prompt generated based on the probability of the next vital slot that needs to be filled in.

User: ⟨*silence*⟩
Dragon Dialogue: *What is the month and day of departure?*

The system produces a more detailed prompt for the slot based on the slot grammar. In other cases the more detailed prompt may be based on a longer path in the domain model.

User: *13th of June.*
Dragon Dialogue: *Let me see if we fly that route.*

Feedback to user. The system checks the backend system. There are two flights that match the user's request.

Dragon Dialogue: *Would you like to leave in the morning or afternoon?*
User: *In the morning.*
Dragon Dialogue: *There is a flight leaving Austin at eight-twenty and arriving Boston at 12:30.*
User: *That sounds good.*

System fills in the estimated time of departure slot and picks the likeliest target action (i.e., price check).

Dragon Dialogue: *Would you need a return booking?*
User: *No.*
Dragon Dialogue: *Would you like to fly coach or first class?*
User: *Coach.*
Dragon Dialogue: *The fare for the booking from Austin to Boston on the 13th of June is five hundred and fifteen dollars. Would you like to make a reservation?*
User: *That's too expensive, I'll call back later.*

Nothing is committed to the backend system.

5   Conclusions

It is widely recognized that the next generation of speech-based user interfaces will involve some dialogue management functionality. We develop techniques through which interactive spoken language systems can be readily constructed by a developer not necessarily well-versed in speech recognition. We realize these techniques as declarative models of the key concepts involved in managing a spoken dialogue. These models are explicitly incorporated as components of our architecture. A related strength of our system is in using the domain models to generate natural language prompts and grammars automatically [Singh *et al.*, 1996].

Future work in designing a portable spoken language dialogue system includes conducting usability testing and evaluation. This includes testing the portability (developer-friendliness) and usability (user-friendliness) of the system. Such testing would include porting the system to new domains by application developers and experiments in user-friendliness. In addition, for more thorough testing, we are hoping to have others apply it in a new domain.

References

[Batini *et al.*, 1992] Batini, Carlo; Ceri, Stefano; and Navathe, Sham; 1992. *Conceptual Database Design: An Entity-Relationship Approach.* Benjamin Cummings.

[Proctor & Young, 1991] Proctor, C. and Young, S.; 1991. Dialogue control in conversational speech interfaces. In Taylor, M.; Neel, F.; and Bouwhuis, D., editors, *The Structure of Multimodal Dialogue.* Elsevier. 385–399.

[Reichman, 1986] Reichman, R.; 1986. *Getting Computers to Talk Like You and Me.* MIT Press.

[Shriber *et al.*, 1992] Shriber, M.; Altom, M.; Macchi, M.; and Wallace, K.; 1992. Human-machine problem solving using spoken language systems (SLS): Factors affecting performance and user satisfaction. In *Proceedings of the DARPA Speech and Natural Language Workshop.*

[Singh & Barnett, 1996] Singh, Mona and Barnett, James; 1996. Designing spoken language interfaces for information retrieval. In *Proceedings of the International Symposium on Spoken Dialogue.* 73–76. Held in conjunction with the International Conference on Spoken Language Processing (ICSLP).

[Singh *et al.*, 1996] Singh, Mona; Barnett, James; and Singh, Munindar; 1996. Generating natural language expressions from entity-relationship diagrams. Unpublished manuscript.

[Tanaka *et al.*, 1996] Tanaka, S.; Nakazato, S.; Hoashi, K.; and Shirai, K.; 1996. Spoken dialogue interface in dual task situation. In *International Symposium of Spoken Dialogue.* 153–156.

[Taylor *et al.*, 1991] Taylor, M. F.; Neel, F.; and Bouwhuis, D.; 1991. *The Structure of Multimodal Dialogue.* Elsevier.

[Waterworth, 1991] Waterworth, J.; 1991. Interactive strategies for conversational computer systems. In Taylor, M.; Neel, F.; and Bouwhuis, D., editors, *The Structure of Multimodal Dialogue.* Elsevier. 331–343.

Minimizing Cumulative Error
in Discourse Context

Yan Qu[1], Barbara Di Eugenio[3], Alon Lavie[2], Lori Levin[2] and Carolyn P. Rosé[1]

[1] Computational Linguistics Program,
[2] Center for Machine Translation,
Carnegie Mellon University, Pittsburgh, PA 15213, USA
[3] Learning Research and Development Center,
University of Pittsburgh, Pittsburgh, PA 15260, USA

Abstract. Cumulative error limits the usefulness of context in applications utilizing contextual information. It is especially a problem in spontaneous speech systems where unexpected input, out-of-domain utterances and missing information are hard to fit into the standard structure of the contextual model. In this paper we discuss how our approaches to recognizing speech acts address the problem of cumulative error. We demonstrate the advantage of the proposed approaches over those that do not address the problem of cumulative error. The experiments are conducted in the context of Enthusiast, a large Spanish-to-English speech-to-speech translation system in the appointment scheduling domain [10, 5, 11].

1 The Cumulative Error Problem

To interpret natural language, it is necessary to take context into account. However, taking context into account can also generate new problems, such as those arising because of cumulative error. Cumulative error is introduced when an incorrect hypothesis is chosen and incorporated into the context, thus providing an inaccurate context from which subsequent context-based predictions are made. For example, in Enthusiast, a large Spanish-to-English speech-to-speech translation system in the appointment scheduling domain [10, 5, 11], we model the discourse context using speech acts to represent the functions of dialogue utterances. Speech act selection is strongly related to the task of determining how the current input utterance relates to the discourse context. When, for instance, a plan-based discourse processor is used to recognize speech acts, the discourse processor computes a chain of inferences for the current input utterance, and attaches it to the current plan tree. The location of the attachment determines which speech act is assigned to the input utterance. Typically an input utterance can be associated with more than one inference chain, representing different possible speech acts which could be performed by the utterance out of context. Focusing heuristics are used to rank the different inference chains and choose the one which attaches most coherently to the discourse context [3, 8]. However, since heuristics can make wrong predictions, the speech act may be misrecognized, thus making the context inaccurate for future context-based predictions.

Unexpected input, disfluencies, out of domain utterances, and missing information add to the frequency of misrecognition in spontaneous speech systems, leaving the discourse processor in an erroneous state which adversely affects the quality of contextual information for processing later information. For example, unexpected input can drastically change the standard flow of speech act sequences in a dialogue. Missing contextual information can make later utterances appear not to fit into the context.

Cumulative error can be a major problem in natural language systems using contextual information. Our previous experiments conducted in the context of the Enthusiast spontaneous speech translation system show that cumulative error can significantly reduce the usefulness of contextual information [6]. For example, we applied context-based predictions from our plan-based discourse processor to the problem of parse disambiguation. Specifically, we combined context-based predictions from the discourse processor with non-context-based predictions produced by the parser module [4] to disambiguate possibly multiple parses provided by the parser for an input utterance. We evaluated two different methods for combining context-based predictions with non-context-based predictions, namely a genetic programming approach and a neural network approach. We observed that in absence of cumulative error, context-based predictions contributed to the task of parse disambiguation. This results in an improvement of 13% with the genetic programming approach and of 2.5% with the neural net approach compared with the parser's non-context-based statistical disambiguation technique. However, cumulative error affected the contribution of contextual information. In the face of cumulative error, the performance decreased by 7.5% for the neural net approach and by 29.5% for the genetic programming approach compared to their respective performances in the absence of cumulative error, thus dragging the performance statistics of the context-based approaches below that of the parser's non-context-based statistical disambiguation technique. The adverse effects of cumulative error in context have been noted in NLP in general. For example, Church and Gale [2] state that "it is important to estimate the context carefully; we have found that poor measures of context are worse than none." However, we are not aware of this issue having been raised in the discourse processing literature.

In the next section, we describe some related work on processing spontaneous dialogues. Section 3 gives a brief description of our system. We discuss the techniques we used to reduce the cumulative error in discourse context for the task of speech act recognition in Section 4. Lastly, we evaluate the effects of the proposed approaches on reducing cumulative error.

2 Related Work

There has been much recent work on building a representation of the discourse context with a plan-based or finite state automaton-based discourse processor

[1, 9, 3, 7, 8, 5]. Of these, the Verbmobil discourse processor [7] and our Enthusiast discourse processor are designed to be used in a wide coverage, large scale, spontaneous speech system. In these systems, the design of the dialogue model, whether plan-based or a finite state machine, is grounded in a corpus study that identifies the standard dialogue act sequences. When the recognized dialogue act is inconsistent with the dialogue model, the systems can rely on a repair procedure to resolve the inconsistency as described in [7].

The Verbmobil repair model [7], however, does not address cumulative error in discourse context. In Verbmobil, every utterance, even if it is not consistent with the dialogue model, is assumed to be a legal dialogue step. The strategy for error recovery, therefore, is based on the hypothesis that the assignment of a dialogue act to a given utterance has been incorrect or rather that the utterance has multiple dialogue act interpretations. The semantic evaluation component in Verbmobil, which computes dialogue act information via the keyword spotter, only provides the most plausible dialogue act. The plan recognizer relies on information provided by the statistical module to find out whether additional interpretations are possible. If an incompatible dialogue act is encountered, the system employs the statistical module to provide an alternative dialogue act, which is most likely to come after the preceding dialogue act and which can be consistently followed by the current dialogue act, thereby gaining an admissible dialogue act sequence. Thus the system corrects context as the dialogue goes along. As we mentioned earlier, contrary to the assumption made in Verbmobil, in spontaneous speech not all utterances fit adequately into the standard dialogue model because of missing information or unexpected input in addition to misrecognition. Moreover, updating context based on the current dialogue state without an evaluation of the current state cannot reduce cumulative error for future predictions and is likely to introduce cumulative error into the context.

3   System Description

Enthusiast is composed of four main modules: speech recognition, parsing, discourse processing, and generation. Each module is domain-independent and language-independent but makes use of domain specific and language specific knowledge sources for customization.

The hypothesis produced by the speech recognizer about what the speaker has said is passed to the parser. The GLR* parser [4] produces a set of one or more meaning representation structures which are then processed by the discourse processor. The output of the parser is a representation of the meaning of the speaker's sentence. Our meaning representation, called an interlingua text (ILT), is a frame-based language-independent meaning representation. The main component of an ILT are the speech act (e.g., **suggest**, **accept**, **reject**), the sentence type (e.g., **state**, **query-if**, **fragment**), and the main semantic frame (e.g., **free**, **meet**). An example of an ILT is shown in Figure 1.

Development of our discourse processing module was based on a corpus of 20 spontaneous Spanish scheduling dialogues containing a total of 630 utterances.

YO PODRÍA MARTES EN LA MAÑANA
(I could meet on Tuesday in the morning)

```
((SENTENCE-TYPE *STATE)
 (FRAME *MEET)
 (SPEECH-ACT *SUGGEST)
 (A-SPEECH-ACT (*MULTIPLE* *SUGGEST *ACCEPT
               *STATE-CONSTRAINT))
 (WHO ((FRAME *I)))
 (WHEN
   ((WH -) (FRAME *SIMPLE-TIME)
         (DAY-OF-WEEK TUESDAY)
         (TIME-OF-DAY MORNING)))
 (ATTITUDE *POSSIBLE))
```

Fig. 1. An Interlingua Text (ILT)

We identify a total of fourteen possible speech acts in the appointment scheduling domain [8] (Figure 2). The discourse processing module disambiguates the speech act of each utterance, updates a dynamic memory of the speakers' schedules, and incorporates the utterance into discourse context.

Speech Act	Example Utterance
Accept	Thursday I'm free the whole day.
Acknowledge	OK, I see.
Address	Wait, Alex.
Closing	See you then.
Confirm	You are busy Sunday, right?
Confirm-Appointment	So Wednesday at 3:00 then?
Deliberation	Hm, Friday in the morning.
Opening	Hi, Cindy.
Reject	Tuesday I have a class.
Request-Clarification	What did you say about Wednesday?
Request-Response	What do you think?
Request-Suggestion	What looks good for you?
State-Constraint	This week looks pretty busy for me.
Suggest	Are you free on the morning of the eighth?

Fig. 2. Speech Acts Covered by Enthusiast

We use four processing components for speech act recognition: a grammar prediction component, a statistical component, a finite state machine, and a plan-based discourse processor. The grammar prediction component assigns a set of possible speech acts to an ILT based on the syntactic and semantic information in the interlingua representation. The resulting set of possible speech acts is inserted into the **a-speech-act** slot of the ILT (See Figure 1). The final determination of the communicative function of the ILT, the speech act, is done by the other three components. The statistical component predicts the following speech act using knowledge about speech act frequencies in our training corpus. The statistical component is able to provide ranked predictions in a fast and efficient way. To cater to the sparse data problem, bigram speech act probabilities are smoothed based on backoff models [12]. The finite state machine (FSM) describes representative sequences of speech acts in the scheduling domain. It is used to record the standard dialogue flow and to check whether the predicted speech act follows idealized dialogue act sequences. The FSM consists of states and transition arcs. The states represent speech acts in the corpus. The transitions between states can have the symbols: S (for the same speaker), C (for change of speaker), or null (no symbol); the null symbol represents the cases in which the transition is legal, independent of whether the speaker changes or remains the same[4]. A graphical representation of the major parts of the FSM appears in Figure 3. We extended the FSM so that at each state of the finite state machine we allow for phenomena that might appear anywhere in a dialogue, such as acknowledge, address, confirm, request-clarification, and deliberation. The plan-based discourse processor handles knowledge-intensive tasks exploiting various knowledge sources, including the grammar component predictions and linguistic information. Details about the plan-based discourse processor can be found in [8]. The finite state machine and the statistical component have recently been implemented as a fast and efficient alternative to the more time-consuming plan-based discourse processor. In our future design of the discourse processing module, we may adopt a layered architecture similar to the one proposed in Verbmobil. In such an architecture, the finite state machine would constitute a lower layer providing an efficient way of recognizing speech acts, while the plan-based discourse processor, at a higher layer, would be used to handle more knowledge intensive processes, such as recognizing doubt or clarification sub-dialogues and robust ellipsis resolution. In this paper, we discuss the cumulative error problem in the context of the finite state machine and the statistical component.

4 Speech Act Recognition

For the task of speech act recognition, we use a combination of grammatical, statistical, and contextual knowledge. The finite state machine encodes the preceding context state, tests the consistency of the incoming utterance with the dialogue model and updates the current state. Given the current state, the finite state machine can provide a set of speech acts that are likely to follow. The speech act of the following input utterance should be a member of this set if the input utterance follows the standard dialogue flow. This set of speech acts is compared with the set of possible speech acts (a-speech-act) proposed by the grammar component for the same input utterance. The intersection of the finite state machine predictions and the grammar component predictions should yield

---

[4] Our corpus analysis showed that certain dialogue act sequences are possible only for the same speaker and others are possible only for different speakers.

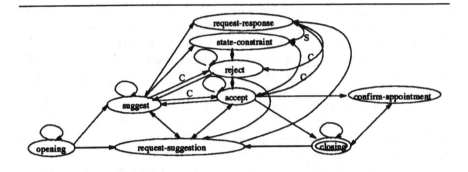

The state **opening** is the initial state. The state **closing** is the final state.
All other states are non-final states.

Fig. 3. The Main Component of the Finite State Machine

the speech acts which are consistent both with the input semantic representation and with the standard dialogue flow. Oftentimes, an utterance can perform more than one legal function. Bigram speech act probabilities are then used to select the most probable one from the intersection set.

An empty intersection between the two sets of predictions signals an inconsistency between the non-context-based grammar predictions and the context-based FSM predictions. The inconsistency can result from unexpected inputs, missing information, out of domain utterances, or simply misrecognized speech act. We tested two approaches for dealing with the conflicting predictions: a *jumping context approach* and a *hypothesis tree approach*. We describe the two approaches below.

Jumping context approach

The rationale behind the jumping context approach is that while we recognize the predictive power of a statistical model, a finite state machine, or a plan-based discourse processor, we abandon the assumption that dialogue act sequences are always ideal in spontaneous speech. Instead of trying to incorporate the current input into the dialogue context, we accept that speech act sequences can at times be imperfect. Instead of following the expectations provided by the context, we assume there is an inaccurate context and there is a need to re-establish the state in the discourse context. In such cases, we trust the grammar predictions more, assessing the current position using syntactic and semantic information. When there is more than one speech act proposed by the grammar component, we use speech act unigrams to choose the most likely one in the corpus. The context state will then be updated accordingly using the grammar prediction. In the graph representation of the finite state machine, this corresponds to allowing empty arc jumps between any two states. Note that this jump from one state to another in the finite state machine is *forced* and *abnormal* in the sense that it is designed to cater to the abrupt change of the flow of dialogue act sequences in spontaneous speech. Thus it is different from transitions with null symbols, which record legal transitions between states. The algorithm for this approach is described in Figure 4. We demonstrate later that this approach gives better performance than one which trusts context in the case of conflicting predictions.

```
context-state = 'start
FOR each input ilt
   context-predictions = predictions from the FSM
                  given context-state
   grammar-predictions = a-speech-act in input ilt
   Intersect context- and grammar-predictions
   IF intersection is not empty,
      use bigrams to rank the speech acts in intersection
      return the most probable follow up speech act
   ELSE ;;; use grammar predictions
      IF more than one speech act in a-speech-act
         use unigrams to rank the possible speech acts
         return the most probable speech act
      ELSE return a-speech-act
   update context-state using the returned speech act
```

Fig. 4. Algorithm for Jumping Context Approach

As an example, consider the dialogue excerpt in Table 1. After speaker S2 accepts S1's suggestion and tries to close the negotiation, S2 realizes that they have not decided on where to meet. The utterance *no* after the closing *chau* does not fit into the dialogue model, since the legal dialogue acts after a **closing** are **closing**, **confirm-appointment** or **request-suggestion** (see Figure 3). When the standard dialogue model is observed (marked by *Strict Context* in Table 1), the utterance *no* is recognized as **closing** since **closing** is the most probable speech act following the previous speech act **closing**. If upon seeing this conflict we instead trust the grammar prediction (marked by *Jumping Context* in Table 1), by recognizing *no* as a **reject**, we bring the dialogue context back to the stage of negotiation. Trusting the grammar, however, does not imply that we should abandon context altogether. In the test set in the experiments discussed in the next section, context represented by the FSM has shown to be effective in reducing the number of possible speech acts assigned by the grammar component.

Hypothesis tree approach

The rationale behind the hypothesis tree approach is that instead of producing a single speech act hypothesis at the time an utterance is passed to the discourse processor, we delay the decision until a later point. In doing so, we hope to reduce cumulative error due to misrecognition because of early commitment to a decision. Specifically, we keep a set of possible speech act hypotheses for

Dialogue Utterances	Strict Context	Jumping Context
S1: QUÉ TE PARECE EL LUNES NUEVE ENTONCES (HOW IS MONDAY THE NINTH FOR YOU THEN)	suggest	suggest
S2: PERFECTO (PERFECT)	accept	accept
CHAU (BYE)	closing	closing
NO (NO)	closing	reject
ESPÉRATE (WAIT)	address	address
NO (NO)	closing	reject
ESPÉRATE (WAIT)	address	address
ALGO PASÓ MAL (SOMETHING BAD HAPPENED)	no parse	no parse
DÓNDE NOS VAMOS A ENCONTRAR (WHERE ARE WE GOING TO MEET)	request-suggestion	request-suggestion
S1: NO (NO)	reject	reject
ESPÉRATE (WAIT)	address	address
SÍ (YES)	acknowledge	acknowledge
DÓNDE NOS ENCONTRAMOS (WHERE ARE WE MEETING)	request-suggestion	request-suggestion

Table 1. An Example Dialogue

each input utterance as contextual states for future predictions. Each context state may in turn be followed by more than one speech act hypothesis for the subsequent utterance, thus yielding a tree of possible sequences of speech act hypotheses. The hypothesis tree is expanded within a beam so that only a certain total number of branches are kept to avoid memory explosion. When the turn shifts between speakers, the hypothesis path with the highest probability (calculated by multiplying speech act bigram probabilities in that path) is chosen as the best hypothesis for the sequences of ILTs in that turn. Each ILT is then updated with its respective speech act in the chosen hypothesis. For each new turn, the last context state in the best hypothesis of the previous turn is used as the starting root for building new hypothesis tree. Figure 5 gives the algorithm for the hypothesis tree approach.

As in the jumping context approach, the predictions of speech acts for each utterance are the combined result of the context-based FSM predictions and non-context-based grammar predictions. The intersection of both predictions gives the possible speech acts which are consistent with both the dialogue model and the default functions of the input utterance. When there is no intersection, we face the decision of trusting the context-based FSM predictions or the non-context-based grammar predictions. We demonstrate later that, for the hypothesis tree approach, again, trusting grammar predictions gives better results than strictly following context predictions at the time of conflicting predictions.

```
hypothesis-tree = '(((start)))
ILTS = nil
FOR each input ilt
  IF still in the same turn
    push ilt into ILTs
    FOR each path in the hypothesis-tree
      context-state = last state in the path
      get speech act predictions for input ilt
      update hypothesis-tree
  ELSE ;;; turn shifts
    choose the path with the highest probability
    update ilts in ILTS with their respective speech act
      prediction in the chosen path
    ILTS = nil
    context-state = last state in the chosen path
    hypothesis-tree = (((context-state)))
    push ilt into ILTS
    get speech act predictions for input ilt
    update hypothesis-tree
  rank the paths in the hypothesis-tree and
  trim the tree within a beam.
```

Fig. 5. Algorithm for the Hypothesis Tree Approach

5 Evaluation

We developed the finite state machine and the statistical module based on the corpus of 20 dialogues mentioned in Section 3. We tested them on another 10 unseen dialogues, with a total of 506 dialogue utterances. Each utterance in both the training and testing dialogues is tagged with a hand-coded target speech act for the utterance. Out of the 506 utterances in the test set, we considered only the 345 utterances that have possible speech acts (in the **a-speech-act** slot) proposed by the non-context-based grammar component.[5]

We conducted two tests on the set of 345 utterances for which the **a-speech-act** slot is not empty. Test 1 was done on a subset of them, consisting of 211 dialogue utterances for which the grammar component returns multiple possible speech acts: we measured how well the different approaches correctly disambiguate the multiple speech acts in the **a-speech-act** slot with respect to the

---

[5] For 161 utterances, the grammar component doesn't return any possible speech act. This is because the parser does not return any parse for these utterances or the utterances are fragments. Although it is possible to assign speech acts to the fragments based on contextual information, we found that, without adequate semantic and prosodic information, the context predictions for these fragments are usually not reliable.

hand-coded target speech act. Test 2 was done on the whole set of 345 utterances, measuring the performance of the different approaches on the overall task of recognizing speech acts.

We evaluate the performance of our proposed approaches, namely the jumping context approach and the hypothesis tree approach, in comparison to an approach in which we always try to incorporate the input utterance into the discourse context (marked by *Strict Context* in Table 2). These approaches are all tested in the face of cumulative error[6]. We also measured the performance of randomly selecting a speech act from the **a-speech-act** slot in the ILT as a baseline method. This method gives the performance statistic when we do not use any contextual information provided by the finite state machine.

Approaches	Test 1	Test 2
Random from Grammar	38.6%	60.6%
Strict Context (Trusting FSM)	52.4%	65.5%
Jumping Context (Trusting Grammar)	55.2%	71.3%
Hypothesis Tree Trusting FSM	48.0%	56.5%
Hypothesis Tree Trusting Grammar	50.0%	60.6%

Table 2. Evaluation: Percent Correct Speech Act Assignments

Table 2 gives some interesting results on the effect of context in spoken discourse processing. Since Test 1 is conducted on utterances with multiple possible speech acts proposed by the non-context-based grammar component, this test evaluates the effects on speech act disambiguation by different context-based approaches. All four approaches employing context perform better than the non-context-based grammar predictions. Test 1 also demonstrates that it is imperative to estimate context carefully. Our experiments show that when context-based predictions and non-context-based predictions are inconsistent with each other, trusting the non-context-based grammar predictions tend to give better results than trusting context-based FSM predictions. In particular, the jumping context approach gives 2.8% improvement over the strict context approach in which context predictions are strictly followed, and trusting grammar predictions gives 2% improvement over trusting FSM predictions in the hypothesis tree approach. To our surprise, the jumping context approach and the strict context approach do better than the hypothesis tree approaches in which more contextual information is available at decision time. This seems to suggest that keeping more contextual information for noisy data, such as spontaneous speech, may actually increase the chances for error propagation, thus making cumulative

---

[6] We found it hard to test in absence of cumulative error. Because of missing information and unexpected input, it is hard even for the human coder to provide an accurate context.

error a more serious problem. In particular, at the point where grammar and context give conflicting predictions, the target speech act may have such a low bigram probability with respect to the given context state that it gives a big penalty to the path of which it is a part.

Test 2 is conducted on utterances with either ambiguous speech acts or unambiguous speech acts proposed by the grammar component. When an ILT has one unambiguous possible speech act, we can assume that the grammar component is highly confident of the speech act hypothesis, based on the syntactic and semantic information available[7]. Note again that the jumping context approach does better than the strict context approach for dealing with conflicting predictions. The hypothesis tree approach, however, does not improve over the non-context-based grammar approach, regardless of whether the grammar predictions are trusted or the context predictions are trusted. This observation seems to support our belief that reestablishing a context state in case of prediction conflicts is an effective approach to reducing cumulative error. Keeping a hypothesis tree to store more contextual information is not as effective as reestablishing the context state, since more contextual information cannot stop error propagation. As decisions are made at a later point, certain target speech acts may be buried in a low probability path and will not be chosen.

6 Conclusion

In this paper we have discussed our effort to minimize the effect of cumulative error in utilizing discourse context. We challenged the traditional assumption that every utterance in a dialogue adheres to the dialogue model and that the process of recognizing speech acts necessarily results in a speech act that can be best incorporated into the dialogue model. We showed that in spontaneous speech, the ideal dialogue flow is often violated by unexpected input, missing information or out of domain utterances in addition to misrecognition. To model dialogue more accurately, this fact should be taken into account. We experimented with two approaches to reducing cumulative error in recognizing speech acts. Both approaches combine knowledge from dialogue context, statistical information, and grammar prediction. In the case of a prediction conflict between the grammar and the context, instead of blindly trusting the predictions from the dialogue context, we trust the non-context-based grammar prediction. Our results demonstrate that reestablishing a context state by trusting grammar predictions in case of prediction conflicts is more robust in the face of cumulative error. Our future work includes exploring different smoothing techniques for the context model in order to quantify the effectiveness of context in different situations.

---

[7] The fact that in 134 utterances there is no speech act ambiguity explains the good performance of the random approach.

7 Acknowledgements

We would like to thank Alex Waibel for discussions and comments over this work and thank Chris Manning, Xiang Tong and the reviewers for references and comments. This work was made possible in part by funding from the U.S. Department of Defense. While this work was carried out, the second author was with the Computational Linguistics Program, Carnegie Mellon University.

References

1. J. F. Allen and L. K. Schubert. *The Trains Project.* University of Rochester, School of Computer Science, 1991. Technical Report 382.
2. K. W. Church and W. A. Gale. Probability Scoring for Spelling Correction. *Statistics and Computing*, 1:93–103, 1991.
3. L. Lambert. *Recognizing Complex Discourse Acts: A Tripartite Plan-Based Model of Dialogue.* PhD thesis, Department of Computer Science, University of Delaware, 1993.
4. A. Lavie. *A Grammar Based Robust Parser For Spontaneous Speech.* PhD thesis, School of Computer Science, Carnegie Mellon University, 1995.
5. L. Levin, O. Glickman, Y. Qu, D. Gates, A. Lavie, C. P. Rosé, C Van Ess-Dykema, and A. Waibel. Using Context in Machine Translation of Spoken Language. In *Theoretical and Methodological Issues in Machine Translation*, 1995.
6. Y. Qu, C. P. Rosé, and B. Di Eugenio. Using Discourse Predictions for Ambiguity Resolution. In *Proceedings of the COLING*, 1996.
7. N. Reithinger and E. Maier. Utilizing Statistical Dialogue Act Processing in Verbmobil. In *Proceedings of the ACL*, 1995.
8. C. P. Rosé, B. Di Eugenio, L. S. Levin, and C. Van Ess-Dykema. Discourse Processing of Dialogues with Multiple Threads. In *Proceedings of the ACL*, 1995.
9. R. W. Smith, D. R. Hipp, and A. W. Biermann. An Architecture for Voice Dialogue Systems Based on Prolog-style Theorem Proving. *Computational Linguistics*, 21(3):218–320, 1995.
10. B. Suhm, L. Levin, N. Coccaro, J. Carbonell, K. Horiguchi, R. Isotani, A. Lavie, L. Mayfield, C. P. Rosé, C. Van-Ess Dykema, and A. Waibel. Speech-Language Integration in a Multi-Lingual Speech Translation System. In *Proceedings of the AAAI Workshop on Integration of Natural Language and Speech Processing*, 1994.
11. A. Lavie, D. Gates, M. Gavaldá, L. Mayfield, A. Waibel and L. Levin. Multilingual Translation of Spontaneously Spoken Language in a Limited Domain. In *Proceedings of the COLING*, 1996.
12. S. M. Katz. Estimation of Probabilities from Sparse Data for the Language Model Component of a Speech Recognizer. In *IEEE Transactions on Acoustics, Speech and Signal Processing*, 1987.

Automatic Evaluation Environment for Spoken Dialogue Systems

Masahiro Araki and Shuji Doshita

Department of Information Science, Graduate School of Engineering,
Kyoto University, Kyoto 606-01 Japan
e-mail: {araki, doshita}@kuis.kyoto-u.ac.jp

Abstract. The need for an evaluation method of spoken dialogue systems as a whole is more critical today than ever before. However, previous evaluation methods are no longer adequate for evaluating interactive dialogue systems. We have designed a new evaluation method that is system-to-system automatic dialogue with linguistic noise. By linguistic noise we simulate speech recognition errors in Spoken Dialogue Systems. Therefore, robustness of language understanding and of dialogue management can be evaluated. We have implemented an evaluation environment for automatic dialogue. We examined the validity of this method for automatic dialogue under different error rates and different dialogue strategies.

1 Introduction

As many Spoken Dialogue Systems (SDSs) are implemented and demonstrated in several research organizations, the need for a total and efficient evaluation method for SDSs is more critical today than ever before. However, as for interactive systems, previous evaluation methods are no longer adequate for evaluating some important points.

For example, the typical subsystem evaluation method that divides SDS into several subsystems in order to evaluate each subsystem independently, cannot measure total robustness of SDS. Also, the Input-Output pair evaluation method, that compares the input (typically speech) with the output (the result of dialogue processing, e.g. database records that satisfy the conditions involved in input), cannot measure interactive ability of SDSs because it uses prerecorded data.

In this paper, we propose a new evaluation method for SDSs, that is system-to-system automatic dialogue with linguistic noise. This automatic evaluation method is the optimal method for measuring the system's ability for problem solving and their robustness. The proposed method is total, effective, and an efficient way for improving the performance of SDSs by tuning parameters easily.

This paper is organized as follows. We give an overview of the SDSs evaluation method in Section 2. We propose a new evaluation environment in Section 3. Then, we show an example of evaluation in Section 4. Finally, we discuss our conclusions and describe the directions for future work in Section 5.

2 Overview of Previous Evaluation Methods

In this section, we present an overview of previous work concerning the evaluation of speech understanding systems, natural language understanding systems and dialogue systems. By surveying these works, we extract the defects of previous evaluation methods that we have to take into consideration in evaluating SDSs.

2.1 Subsystem evaluation method

Previously, most SDSs were evaluated by the subsystem evaluation method. According to this method, an SDS is divided into subsystems in order to evaluate each subsystem independently (Figure 1).

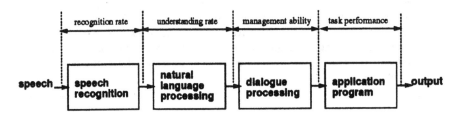

Fig. 1. Concept of subsystem evaluation method

Concerning speech recognition subsystems, well established evaluation methods are word recognition rate or sentence recognition rate. Also in speech recognition subsystems, the difficulty of the target task domain is measured in terms of perplexity.

Concerning language processing subsystems, the developments of task independent evaluation methods are now in progress. One of these works is SemEval (Moore 1994). In SemEval, the meaning of a sentence is represented by predicate-argument structures, which provide a semantically-based and application-independent measure.

However, the subsystem evaluation method cannot capture the cooperation between subsystems. Some kinds of speech recognition error can be recovered using linguistic knowledge. Also some kinds of syntactic / semantic ambiguity can be resolved using contextual knowledge. The ability of dealing with such problems, that is *robustness*, is obtained through the cooperation of subsystems. But the subsystem evaluation method ignores the possibility of these cooperations. Therefore, the subsystem evaluation method is inadequate. Total evaluation is essential for SDSs.

2.2 Input-Output pair evaluation

In the ATIS (Air Traffic Information Service) domain, spoken dialogue systems are evaluated by language input / database answer pairs (Hirshman 1994). This allows us to evaluate total understanding in terms of getting the right answer for a specific task (Figure 2).

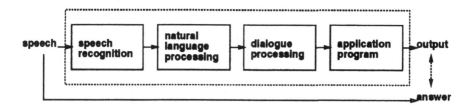

Fig. 2. Concept of Input-Output pair evaluation

However, such evaluation methods cannot measure the interactive ability of SDSs, because they use prerecorded data. For evaluating interactive systems, prerecorded data cannot be used because the user's response determines what the system does next. The system's ability for interactive problem solving or for recovering from miscommunication cannot be evaluated by such methods. Therefore, we must enlarge the scope of evaluation still further to include interactive aspects of SDSs.

2.3 Evaluation by human judges

The alternative for evaluating SDSs uses human judges (Figure 3). The evaluation is made by the task completion rate / time and by a questionnaire filled out by the human subjects.

This method is vital at the very last stage of evaluation of consumer products. But once evaluation includes human factors, it looses objectivity. Also human judgements take much time and are costly. At the early stage of research system development, we need more quick and low cost evaluation methods.

2.4 System-to-system automatic dialogue

A promising way of interactive system evaluation is through system-to-system automatic dialogue (Figure 4) (Carletta 1992; Walker 1994a, 1994b; Hashida et al. 1995). Both input and output of each system are natural / artificial language texts. Dialogue is mediated by a coordinator program. The coordinator program opens a communication channel at the beginning of dialogue, and closes it at

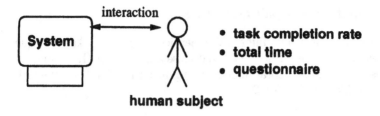

Fig. 3. Concept of evaluation by human judges

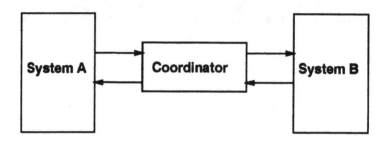

Fig. 4. Concept of system-to-system automatic dialogue

the end of the dialogue. Also, the coordinator program records each utterance and judges whether the task has been successfully completed.

In Japan, a system-to-system dialogue competition, *DiaLeague*, has taken place (Hashida et al. 1995). The task of this competition is shortest route search under incomplete and different route maps. The maps are similar to railway route maps. The map consists of stations and connections. But some connections between stations are inconsistent in the individual maps. Each system must find out the common shortest route from start station to goal station. Each system exchanges information about the map by using natural language text.

The purpose of this competition is to measure the system's ability of problem solving and the conciseness of the dialogue. In (Hashida et al. 1995), they defined the ability of problem solving as the task completion rate. Also, they defined conciseness of dialogue as the number of content words. A smaller number of content words is preferable.

Such an automatic evaluation method can measure total performance and the interactive aspect of a dialogue system. Also, it is easy to test repeatedly. But this is for the dialogue system using written text. We extend this evaluation method for SDSs in the next section.

3 Total and Interactive Evaluation of Spoken Dialogue Systems

In this section, we describe our evaluation method for SDSs. First, we explain the concept of our method. Second, we describe an evaluation environment for system-to-system automatic dialogue with linguistic noise. Next, in order to make this evaluation independent of task, we define the concept of flexibility of an utterance and the flexibility of a dialogue. Finally, we discuss system parameters concerning the dialogue strategy.

3.1 System-to-system dialogue with linguistic noise

We have extended the concept of system-to-system automatic dialogue for the evaluation of SDSs (Figure 5). Random linguistic noise is put into the communication channel by the dialogue coordinator program. This noise is designed for simulating speech recognition errors. The point of this evaluation is to judge the subsystems' ability to repair or manage such misrecognized sentences by a robust linguistic processor or by the dialogue management strategy.

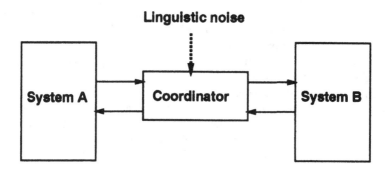

Fig. 5. Concept of system-to-system dialogue with linguistic noise

With our method, the performance of a system is measured by the task achievement rate (ability of problem solving) and by the average number of turns needed for task completion (conciseness of dialogue) under a given recognition error rate.

3.2 Automatic dialogue environment

We have implemented an environment for the evaluation of automatic system-to-system dialogues. Figure 6 shows the concept of the environment.

The environment consists of one coordinator agent and two dialogue agents. At the start of the dialogue, the coordinator sends a start signal to one of the dialogue agents. The dialogue agent who receives the start signal (system A)

```
┌─────────────────────────────────────────────────────────────┐
│              System to System Automatic Dialogue             │
│  ┌─────────────────────────────┐ ┌───────────────────────┐   │
│  │       Dialogue Record       │ │        System A       │   │
│  │  A: Go to 'a' station.      │ │  Go to 'a' station.   │   │
│  │  B: Yes.                    │ │  Go to 'b' station.   │   │
│  │  A: Go to 'b' station.      │ │                       │   │
│  │  B: Where is 'b' station?   │ │                       │   │
│  └─────────────────────────────┘ └───────────────────────┘   │
│  ┌─────────────────────────────┐ ┌───────────────────────┐   │
│  │         Coordinator         │ │        System B       │   │
│  │  Error rate  [  10  ]%      │ │  Yes.                 │   │
│  │   ○  insert                 │ │  Where is 'b' station?│   │
│  │   ●  delete                 │ │                       │   │
│  │   ○  substitute             │ │                       │   │
│  └─────────────────────────────┘ └───────────────────────┘   │
└─────────────────────────────────────────────────────────────┘
```

Fig. 6. Concept of Evaluation environment

opens the dialogue. System A generates natural language text which is sent to the coordinator. The coordinator receives the text, puts linguistic noise into it at the given rate, and passes the result to another dialogue agent (system B). System B has the next turn. The dialogue ends when one of the dialogue agents cuts the connection or when the number of turns exceeds the given upper bound.

The result of the dialogue is examined using logged data. In case both agents reach the same and correct answer, we regard the task problem as solved. The task achievement rate is calculated from the number of dialogues that reach the same and correct answer divided by the total number of dialogues. In addition, we assume that the conciseness of a dialogue can be measured by the average number of turns. This is because SDS puts a strain on the user each time he has to produce an utterance. Therefore we think that fewer turns is preferable.

To make these values independent of task, we defined the flexibility of an utterance and the flexibility of a dialogue which we will describe in the following subsection.

3.3 Flexibility of utterance and dialogue

In order to make our evaluation method independent of task, we think another viewpoint must be added. In SDSs, language processing subsystems must deal with illegal input, such as ungrammatical sentences, sentences with unknown words, speech recognition errors, etc. However, if the correspondence between the sentence and its semantic representation is simple, then it is easy to recover errors or to collect partial results. In this situation, it needs fewer extra turns for the error recovery. As a result, the average number of turns is largely affected by the complexity of the correspondence between the sentence and its semantic representation.

The same is true concerning dialogue management subsystems. A simple dialogue structure can reduce the extra turns for the error recovery.

In order to measure the complexities of these elements, we define for each task a *distance* from an input utterance to its target semantic representation. We call this the *flexibility of an utterance*. The flexibility of an utterance is based on the distance from the result of the speech recognizer to the corresponding predicate-argument structure. The predicate-argument structure is a kind of frame representation of the sentence meaning. The main verb of the sentence always determines the frame name, that is, the predicate. Nouns or noun phrases are used to fill the value slots of arguments.

For defining the *flexibility of an utterance*, we have to specify the input for the language processing subsystems. Because the system deals with spontaneous speech, we can assume a word lattice as input.

From the viewpoint of structural complexity of predicate-argument structure, we define the rank of flexibility of an utterance as follows:

1. **Ordered word sequence**

   An ordered word sequence is characterized by its content words and their order. An ordered word sequence corresponds to exactly one semantic representation (Figure 7). Typically, one of the content words corresponds to the slot of the verb in the predicate-argument structure, the rest of the content words simply fill the value slots of the arguments. Every word in a sequence has, in the given task, only one meaning entry in the dictionary. By this constraint, even if the speech recognizer outputs only content words and their order (e.g. by using word spotter), language processor can decide the proper slot of the words.

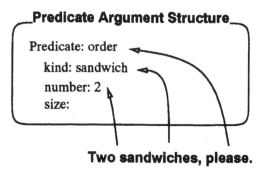

Fig. 7. An example of ordered word sequence

2. **Simple sentence**

   A simple sentence is defined as a type of sentences the semantic representation of which has only one predicate (Figure 8). Obviously, the ordered word sequence rank is a subset of the simple sentence rank. In a simple sentence, some content words can occupy a couple of value slots of the predicate-argument structure. Therefore, in contrast with the ordered word sequence, it needs structural information for the utterance in assigning the value of an argument. A possible parsing method used on this rank is keyword-driven parsing (Araki et al. 1993).

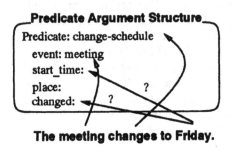

Fig. 8. An example of simple sentence

3. Complex sentence

A Complex sentence is defined as a type of sentence whose argument value can also be a predicate argument structure (Figure 9). This definition is almost the same as in the linguistic terminology. It needs more structural information of the utterance in assigning the value of an argument than does the simple sentence, because the possible value slots are increased in the predicate-argument structure.

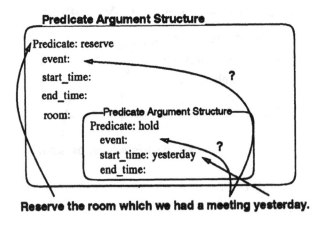

Fig. 9. An example of complex sentence

However, if there exist tight dialogue level constraints, the *distance* seems to be diminished. We also define the *flexibility of a dialogue*. Considering the influence on language processing, we define ranks by using the following notion of dialogue model:

1. **Automaton dialogue model**
   Automaton dialogue models are in the class of regular grammars. The model strictly limits the flexibility of a dialogue with state transitions. Because of this limitation, expectations regarding the next utterance can be used powerfully.

2. **Plan recognition based dialogue model**
   Plan recognition based dialogue models are in the class of context free grammars. For example, this model is implemented by an event hierarchy (Kautz 1990; Vilain 1990). The advantage of this rank of dialogue model is the tractability of focus shift which is limited in automaton dialogue models.

3. **Dialogue model in different knowledge**
   Dialogue models in different knowledge are proposed by Pollack (1990) and also Grosz et al. (1990). Different knowledge means that participants of the dialogue do not share the same knowledge about plans and actions in the task domain. These models require the modeling of the user's mental state. Some studies employ first order predicate logic for representing the mental state. In general, the computational cost of this rank of model is higher than the other dialogue models. Also, the framework of this rank of dialogue model is mainly in tracing the change of the mental state. Such a framework is not suitable for the prediction of the next utterance.

3.4 Parameters of a dialogue strategy

We think that the dialogue strategy is another important factor for evaluating spoken dialogue systems. What types of feedback or what error recovery techniques are suitable using a given recognition error rate? What level of initiative is suitable for a given situation? These factors should be examined by overall system evaluation.

In our method, a dialogue strategy is represented by parameters of dialogue systems (e.g. level of initiative, type of feedback, frequency of confirmation, etc.). By changing these parameters, most suitable settings of parameters can be discovered in this evaluation environment.

4 Example of Evaluation by Automatic Dialogue

In this section, we show an example of an overall system evaluation. We also discuss the validity of this evaluation method.

4.1 Purpose

In this example, we try to acquire the following information through examining automatic dialogue.

1. How does the number of turns in a dialogue depend on the number of speech recognition errors?
2. What type of dialogue strategy is most appropriate given a specific level of recognition accuracy?

4.2 Conditions

As the task domain for this experiment we selected shortest route search. This is identical with the *DiaLeague* task. Each dialogue system has a route map of a railroad, which is somewhat different from the map the other system works with. Examples of a simplified map are shown in Figure 10. Some line might be cut, or some station name might be eliminated. The purpose of this task is to find out a shortest common route from start station to goal station.

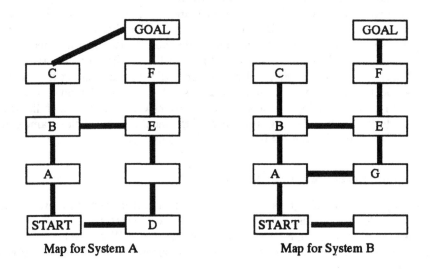

Map for System A   Map for System B

Fig. 10. Examples of maps

In this experiment, we limited the flexibility of an utterance, ordered word sequence rank. The flexibility of dialogue was determined as automaton rank. Of course, a higher rank can be employed for this task. But examining the data of simulated human-human dialogue, we decided to model this task at the lowest rank of utterance and dialogue.

The employed automaton had 8 states and the number of sentence type was 9. Total words in this task was about 70. Among these words, 44 words were the name of stations in the map.

In this experiment, we set the error rate of speech recognition 10 % / 30 %. We simulate speech recognition errors by dropping one content word at the given rate. This type of error reflects the errors often occurring in the case of template matching in robust parsing.

We prepared two dialogue strategies: ask back type and infer type. The ask back strategy means that in case the system detects a speech recognition error, it asks again using the utterance "*Mou ichido itte kudasai*. (I beg your pardon?)" On the other hand, the infer type strategy means that the system tries to infer what has to be said. The recovery strategy is that if a verb is missing, then the system infers the verb from the rest of the content words; if an important noun is left out, then the system gives up recovering and asks again. Multiplying these two conditions, we get four types of dialogue conditions. For each condition, we made three trials using the same map.

4.3 Results

In all 12 cases the goal is reached because of a simple rank of utterance and dialogue. Therefore, the task achievement rate is 100 %. Also, we counted the number of turns and calculated average turns to measure the conciseness of the dialogue. Table 1 shows the results of this experiment.

Table 1. Average number of turns under different conditions

error rate(%)	dialogue strategy	average number of turns
0	—	102
10	ask back	112
10	infer	113
30	ask back	149
30	infer	123

The first column shows that it takes 102 turns to solve the problem if there are no recognition errors. It serves as a baseline for other experiments. The second column shows the results of examining the availabilities of two types of dialogue strategy respectively. They are achieved under 10 % recognition errors. It yields about 10 % of redundant interactions by recognition errors in both types. Even using different dialogue strategies, there is little difference in the average number of turns. The third column shows the results under 30 % recognition errors. Another 33 % of redundant interactions were yielded by the ask back strategy, but 9 % from the infer strategy.

4.4 Discussion

¿From this experiment, using ordered word sequence rank and automaton rank, we can say that the infer type strategy is not effective at relatively low recognition rates. But if the recognition rate is high, the infer type strategy is more effective than the ask back strategy.

It seems a natural conclusion. However, we think that this conclusion shows the validity of our evaluation method.

5 Conclusion and Future Research

We proposed a new evaluation method and its environment for spoken dialogue systems by system-to-system automatic dialogue with linguistic noise. Also, we defined the rank of an utterance and the rank of a dialogue in order to make the evaluation independent of task. The validity of this evaluation method is shown by a sample experiment.

The remaining problems are (1) to make a good linguistic generator for noise, and (2) to establish a measure of difficulty for each utterance and dialogue rank, which is similar to the perplexity measure in speech recognition.

In future research, we intend to implement another dialogue system, such as Araki et al. (1995), that covers all the combinations of utterance and dialogue ranks.

Acknowledge

We would like to thank the reviewers for their comments which helped to improve this paper.

References

1. Araki, M., Kawahara, T. and Doshita, S.: A keyword-driven parser for spontaneous speech understanding. In *Proc. Int'l Sympo. on Spoken Dialogue* (1993) 113–116
2. Araki, M. and Doshita, S.: Cooperative spoken dialogue model using Bayesian network and event hierarchy. *Trans. of IEICE*, E78-d(6) (1995) 629–635
3. Carletta, J. C.: *Risk-taking and Recovery in Task-Oriented Dialogue*. PhD thesis, University of Edinburgh, (1992)
4. Grosz, B. J. and Sidner, C. L.: Plans for discourse. In Cohen, P. R., Morgan, J. and Pollack, M. E. editors, *Intentions in Communication*. The MIT Press, (1990) 417–444
5. Hashida, K. et al.: DiaLeague. In *Proc. of the first annual meeting of the association for natural language processing (in Japanese)* (1995) 309–312
6. Hirshman L.: Human language evaluation. In *Proc. of ARPA Human Language Technology Workshop* (1994) 99–101
7. Kautz, H. A.: A circumscriptive theory of plan recognition. In Cohen, P. R., Morgan, J. and Pollack, M. E. editors, *Intentions in Communication*. The MIT Press, (1990) 105–133
8. Moore, R. C.: Semantic evaluation for spoken-language systems. In *Proc. of ARPA Human Language Technology Workshop* (1994) 126–131
9. Pollack, M. E.: Plans as complex mental attitudes. In P. R. Cohen, J. Morgan, and M. E. Pollack, editors, *Intentions in Communication*. The MIT Press, (1990) 77–103
10. Vilain, M.: Getting serious about parsing plans: a grammatical analysis of plan recognition. In *Proc. of AAAI* (1990) 190–197
11. Walker, M. A.: Discourse and deliberation: Testing a collaborative strategy. In *Proc. of COLING94* (1994) 1205–1211
12. Walker, M. A.: Experimentally evaluating communicative strategies: The effect of the task. In *Proc. of AAAI94* (1994) 86–93

End-to-End Evaluation in JANUS:
A Speech-to-Speech Translation System

Donna Gates[1], Alon Lavie[1], Lori Levin[1],
Alex Waibel[1-2], Marsal Gavaldà[1],
Laura Mayfield[1], Monika Woszczyna[2]
and Puming Zhan[1]

[1] Center for Machine Translation,
Carnegie Mellon University,
5000 Forbes Ave.,
Pittsburgh, PA 15213, USA
[2] Universität Karlsruhe,
Fakultät für Informatik,
76131 Karlsruhe, Germany

Abstract. JANUS is a multi-lingual speech-to-speech translation system designed to facilitate communication between two parties engaged in a spontaneous conversation in a limited domain. In this paper we describe our methodology for evaluating translation performance. Our current focus is on *end-to-end evaluations* - the evaluation of the translation capabilities of the system as a whole. The main goal of our end-to-end evaluation procedure is to determine translation accuracy on a test set of previously unseen dialogues. Other goals include evaluating the effectiveness of the system in conveying domain-relevant information and in detecting and dealing appropriately with utterances (or portions of utterances) that are out-of-domain. End-to-end evaluations are performed in order to verify the general coverage of our knowledge sources, guide our development efforts, and to track our improvement over time. We discuss our evaluation procedures, the criteria used for assigning scores to translations produced by the system, and the tools developed for performing this task. Recent Spanish-to-English performance evaluation results are presented as an example.

1 Introduction

JANUS [8, 9] is a multi-lingual speech-to-speech translation system designed to facilitate communication between two parties engaged in a spontaneous conversation in a limited domain. In this paper we describe our methodology for evaluating the translation performance of our system. Although we occasionally evaluate the performance of individual components of our system, our current focus is on *end-to-end evaluations* - the evaluation of the translation capabilities of the system as a whole [1]. Translation in JANUS is performed on basic semantic dialogue units (SDUs). We thus evaluate translation performance on this level. SDUs generally correspond to a semantically coherent segmentation of an utterance into speech-acts.

The main goal of the end-to-end evaluation procedure is to determine the translation accuracy of each of the SDUs in a test set of unseen utterances. The utterances are taken from recorded dialogues in which two speakers schedule a meeting. The end-to-end evaluations on unseen data are performed in order to verify the general coverage of our lexicons, grammars and semantic representations. The evaluations guide our development efforts, and allow us to track our improvement over time. Because our system is designed for the limited domain of scheduling, we are also interested in evaluating the effectiveness of the system in conveying domain-relevant information and in detecting and dealing appropriately with utterances (or portions of utterances) that are out-of-domain. Detection of out-of-domain material allows the system to recognize its own limitations and avoid conveying false or inaccurate information.

JANUS is evaluated on recordings and transcriptions of human-human scheduling dialogues. Test sets for evaluations are always taken from a reserved set of completely "unseen" dialogues. A test set is considered fully unseen only if the speakers of the dialogues have not been used for training the speech recognizer, and the dialogues themselves have not been used for development of the translation components. The performance results reported in this paper were conducted on such test sets. We strongly believe that this method of evaluation is the most meaningful and realistic register of performance of a speech translation system.

The remainder of the paper is organized in the following way. Section 2 presents a general overview of our system and its components. Section 3 contains a detailed description of the evaluation procedure, the criteria used for assigning scores to translations, and the tools developed for performing this task. In Section 4, as an example of these procedures at work, we present the results of recent Spanish-to-English performance evaluations. Finally, a summary and conclusions are presented in Section 5.

2   System Overview

The JANUS system is composed of three main components: a speech recognizer, a machine translation (MT) module and a speech synthesis module. A diagram of the general architecture of the system is shown in Figure 1. The speech recognition component of the system is described elsewhere [11]. For speech synthesis, we use a commercially available speech synthesizer.

The MT module is composed of two separate translation sub-modules which operate independently. The first is the Generalized LR (GLR) module [4, 3], designed to be more accurate. The second is the Phoenix module [6], designed to be more robust. Both modules follow an interlingua-based approach. The source language input string is first analyzed by a parser. In the case of the GLR module, lexical analysis is provided by a morphological analyzer [5, 2]. Each parser produces a language-independent interlingua text representation. The interlingua is then passed to a generation component, which produces a target language output string.

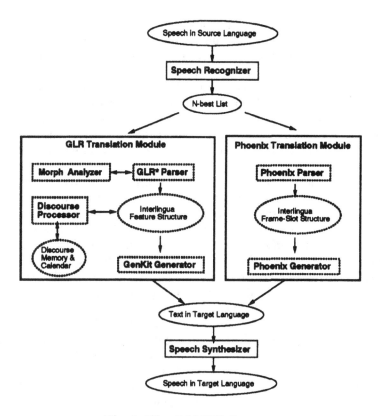

Fig. 1. The JANUS System

Both translation modules are equipped with procedures for detecting parts of utterances that are outside of the scheduling domain. Our goal here is to avoid partial translations of out-of-domain SDUs that force misleading interpretations of the input SDU. For example, we wish to avoid a situation in which *"Tengo dos hijos" (I have two children)* may be translated as *"I have two o'clock free"* This can happen because the out-of-domain word *hijos* is skipped during analysis.

The discourse processor is a component of the GLR translation module. The discourse processor disambiguates the speech act of each sentence, normalizes temporal expressions, and incorporates the sentence into a discourse plan tree. The discourse processor also updates a calendar which keeps track of what the speakers have said about their schedules. The discourse processor is described in detail elsewhere [7].

3   The End-to-end Evaluation Procedure

In order to assess the overall effectiveness of our two translation modules in a consistent and cost effective manner, we developed a detailed end-to-end evaluation procedure. Prior evaluation procedures involved a cumbersome process of comparing the output of individual system components to hand-coded output. In the initial phase of development, this method was found to be very useful. However, it proved to be too labor intensive to hand-code output and evaluate each component of the system when frequent evaluations were needed. In addition, the two translation modules required separate and very different procedures for evaluating their internal components. The solution was to use end-to-end translation output as the basis for the evaluation. This became possible only after the translation modules were fully integrated with the speech recognition component.

The development/evaluation cycle of our translation modules proceeds in the following way. System development and evaluation are performed on batches of data, each consisting of roughly 100 utterances. We refer to these batches as *test sets*. The test sets are chosen from a large pool of "unseen" data reserved for evaluations. System performance on each test set is first evaluated prior to any development based on the data. This allows us to isolate utterances (or parts of utterances) that are not translated adequately by the translation modules. Our translation development staff then augments the analysis and generation knowledge sources in order to improve their coverage, guided by the set of poorly translated examples that were isolated from the test set. Following development, the test sets are re-processed through the system using the updated knowledge sources. They are then re-scored, in order to measure the effect of the development on system performance. Once we are satisfied with the performance level on the current data set, we proceed to a new "unseen" test set and begin the process again. After post-evaluation development, we backup the current version of each translation module.

We believe evaluations should always be performed on data that is used neither for training speech recognition nor for developing the translation modules. All results reported in this paper are based on this type of evaluation. At the end of a development/evaluation cycle, we find that when we retest the data, for transcribed data we typically achieve over 90% correct translations. However, we believe that testing on unseen data before development represents a more valid test of how the system would perform.

Evaluation is normally done in a "batch" mode, where an entire set of utterances is first recognized by the speech recognizer and then translated by the translation modules. Recorded speech for each utterance is processed through the speech recognizer, and an output file of the top recognition hypothesis for each utterance is produced. This file is then passed on to the translation modules, where the utterances are translated and the translation output is saved in an output file. Additionally, a file containing human transcribed versions of the input utterances is also processed through the machine translation modules. Both translation output files are then evaluated. The evaluation of transcribed input allows us to assess how well our translation modules would function with "perfect" speech recognition.

At least once a year we perform large scale system evaluations. The goals of the large evaluation are to measure the current performance of the system and to measure the progress made in development over a specific period of time. To measure progress over time, we take several backed up versions of the translators from significantly different points in time (at least 4 months apart) and run each of them over the same set of unseen test data. The translations are then scored and the end result is a series of scores that should increase from the oldest version to the most recent version.

3.1 Scoring Utterances

When scoring utterances we find that the most accurate results are derived by subdividing a spoken utterance into coherent semantically based chunks. Since an utterance in spontaneous speech may be very short or very long, we assign more than one grade to it based on the number of sentences or fragments it contains. We call these sentences or fragments "semantic dialogue units" (or SDUs). The utterance is broken down into its component SDUs in order to give more weight to longer utterances, and so that utterances containing SDUs in multiple semantic domains can be judged more accurately. Each SDU translation is assigned one grade.

Transcribed data contains markers that represent the end of an SDU. These markers are encoded by hand with the symbol "{seos}" ("semantic end of segment") as shown in the utterance: *sí {seos} está bien {seos} qué día te conviene más a ti {seos}* (English: "yes" "it's ok" "what day is more convenient for you"). Since this example has three SDUs, it will receive three grades.

Translations of speech recognizer output are scored by comparing them to the transcribed source language text. When scoring the translations from the output of speech recognition, the number of grades per utterance is determined by the number of SDUs in the transcribed source language dialogue. Since the output of speech recognition does not contain the SDU markings, the scorer is required to align the recognition output to the transcribed source language dialogue by hand. Then, for each SDU in the transcribed source language dialogue, the scorer must determine whether the translation of the speech recognition output is correct. This method also allows us to determine whether or not a mistranslation is due to an error in speech recognition.

3.2 Grading Criteria

When assigning grades to an utterance the scorer must make judgements as to the relevance of the SDU to the current domain, the acceptability of the translation as part of the meeting scheduling task, and the quality and fluency of acceptable translations. Figure 2 illustrates the decision making process required to arrive at the various translation grades. Letters that appear in parentheses under the grade category correspond to the letter grade assigned by the scorer when using the grading assistant program described in Subsection 3.3.

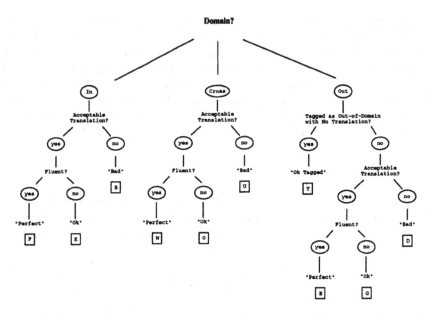

Fig. 2. Decision Making Process for Grading an SDU

Determining Domain Relevance The grader first classifies each utterance as either relevant to the scheduling domain (in-domain), relevant to all domains (cross-domain) or not relevant to the scheduling domain (out-of-domain). We established the following criteria for determining the relevance of each SDU to our domain. In-domain SDUs contain information that will be used specifically for scheduling a meeting such as suggesting a meeting or time, confirming a meeting or time, declining a suggested time or stating a scheduling constraint. If the SDU contains none of this information and is a typical greeting or dialogue maintenance phrase, then the SDU is cross-domain (e.g., *Hola* "hello" or *Pues* "well...") If the SDU contains no scheduling information, does not imply a scheduling restriction or suggestion, and is not a dialogue maintenance phrase, then the SDU is out-of-domain (e.g., discussing one's children). Some SDUs can only be assigned a domain grade based on the context in which it appears. This means that the SDU *Qué tal* which means "How are you?" or "How's that?" may be considered in-domain or cross-domain. If it used in the context of a greeting "Hello... How are you?", then we consider it cross-domain. If it is used in the context of suggesting a time "I can meet on Tuesday at two. How's that?", then we consider it in-domain. Likewise, the SDU *Tengo dos* which means "I have two" may be considered in-domain if preceded by an utterance such as "How many mornings do you have free this week?" or it may be considered out-of-domain if preceded by an utterance such as "So, how many children do you have?". The responses "Yes" and "No" behave similarly and may fall into any of the three domain categories depending on their context.

Perfect	Fluent translation with all information conveyed
OK	All important information translated correctly but some unimportant details missing or translation is awkward
OK tagged	The sentence or clause is out-of-domain and no translation is given.
Bad	Unacceptable translation

Fig. 3. Evaluation Grade Categories

Determining Translation Quality After the domain of the SDU has been determined, the scorer then proceeds to assign one of the translation quality-accuracy grades listed in Figure 3. The grades "Perfect", "OK" and "Bad" are used for judging in-domain, cross-domain and out-of-domain SDUs. "OK Tagged" is used when the translation system correctly recognizes the SDU as out-of-domain and does not translate it.

When a translation is judged as accurately conveying the meaning of the SDU, it is assigned the grade of "Perfect" or "OK". The grade "Perfect" is assigned to a high quality translation that contains all of the information of the input SDU and is presented in a fluent manner, provided the input was also fluent. For example *Tengo un almuerzo con Pedro a las dos* receives the grade "Perfect" when translated as "I have a lunch date with Pedro at two" or "I am having lunch with Pedro at two." The grade "OK" is assigned when the translation is awkward or missing some non-essential information. For example *Tengo un almuerzo con Pedro a las dos* receives the grade "OK" when translated as "I have a lunch date at two" In this example, *Pedro* is not considered crucial for conveying that the speaker is busy at two o'clock. In reporting our results we use the category "acceptable" to represent the sum of the number of "Perfect" and "OK" translations. In addition to these two grades, out-of-domain SDUs may be assigned the grade "OK Tagged" as explained above. The "OK tagged" SDUs are included in the "acceptable" category for out-of-domain translations. A "Bad" translation is simply one that is not acceptable.

One of the drawbacks of relying on human judges to score translations is their subjectiveness. We find that scores from different judges may vary by as much as 10 percentage points. In addition, system developers do not make ideal judges because they are naturally biased. We believe that the most reliable results are derived from employing a panel of at least three judges who are not involved in system development to score the translations. Their scores are then averaged together to form the final result. When the only judges available are people who work on system development, it is absolutely necessary to cross grade the translations and average the results. When one judge is used, he or she cannot be affiliated directly with development.

3.3 The Grading Assistant Program

To assist the scorers in assigning grades to the utterances, we have a simple program that displays dialogues and translations; prompts the scorer for grades; saves, tabulates and averages results; and displays these results in a table. When the same dialogue is translated by both translation modules, or by different versions of the same module, the grading program allows the scorer to compare the two translations, copying grades where the output is identical. When utterances are particularly long, the scorer may loop through the SDUs showing each transcribed SDU with its translation and assign it a grade. The program encourages the user to assign the same number of grades to an utterance as it has SDUs. The symbols representing the possible grades are defined in Figure 4. These grade symbols are used to assign both the quality grade and the domain relevance to an SDU at the same time.

```
p  perfect in-domain
w  perfect cross-domain
e  perfect out-of-domain
k  ok in-domain
o  ok cross-domain
g  ok out-of-domain
t  ok out-of-domain tag (not translated)
b  bad in-domain
u  bad cross-domain
d  bad out-of-domain
```

Fig. 4. Grades Used with the Grading Assistant Program

Figure 5 shows an example of a translation displayed by the grading assistant program with grades assigned to it by a scorer. The first SDU is scored as perfectly translated and cross-domain (W). The meaning of this SDU in context implies that the speaker is merely acknowledging understanding of the previous utterance rather than accepting the suggestion. We consider these dialogue management expressions to be cross-domain. The second SDU "no" is a direct response to an in-domain suggestion so it is assigned an in-domain grade. The third SDU is clearly an in-domain SDU. Both of these translations were considered to be perfect (P) by the scorer.

4 The Evaluation Results

Evaluations are performed periodically to assess our current translation performance and our progress over time. Because speech data often varies in style and content, performance on different evaluation sets may vary significantly. To compensate for this variability, we track progress over time by testing different versions of the translation system on the same set of unseen data. We conduct

11 (fbcg_04_11)

Transcribed:

((+s+ okay) (no) (yo tengo una reunio1n de diez a once))

Generated:

("okay" "no" "i have a meeting from ten o'clock to eleven o'clock")

(# SDUs = 3)

Grade: (w p p)

Fig. 5. Example Display of a Translation with the Grading Assistant Program

In Domain (240 SDUs)

	GLR*		Phoenix	
	transcribed	speech 1st-best	transcribed	speech 1st-best
Perfect	69.6	45.2	51.7	28.7
OK	13.3	8.9	24.6	19.8
Bad	17.1	21.0	23.8	30.8
Recog Err	0.0	25.0	0.0	20.7
Out of Domain (64 SDUs)				
Perfect	54.7	28.1	37.5	14.1
OK	6.3	4.7	9.4	6.3
OK Tagged	25.0	34.4	43.8	43.8
Bad	14.1	7.8	9.4	9.4
Recog Err	0.0	25.0	0.0	26.6
Acceptable Task Performance (Perfect + OK + OK Tagged)				
In Dom	82.9	54.0	76.3	48.6
Out of Dom	85.9	67.2	90.6	64.2
All Dom	83.6	56.7	79.3	51.8
Acceptable Translation Performance (Perfect + OK)				
In Dom	82.9	54.0	76.3	48.6
Out of Dom	61.0	32.8	46.9	20.4
All Dom	78.3	49.5	70.1	42.7

Fig. 6. Percentages for April 1996 Evaluation of GLR* and Phoenix on Three Dialogues

similar evaluations with other source and target languages (English, German, Korean and Japanese). Recent performance results for these languages appear in [10].

The results in Figure 6 were obtained from our April 1996 Spanish-to-English evaluation. The scoring for this evaluation set was conducted prior to the introduction of the cross-domain category. The cross-domain SDUs are included in the out-of-domain category. Figure 6 shows a breakdown of the evaluation results for 3 unseen Spanish dialogues containing 103 utterances translated into English by the GLR* and Phoenix translation modules. The translations were scored by an independent scorer not involved in the development of the system. The independent scorer was a bilingual student fluent in Spanish and English with no formal training in linguistics or translation.

Acceptable translation performance is measured by summing the "Perfect" and "OK" scores. Acceptable task performance for the scheduling domain is measured by summing the "Perfect", "OK" and "OK tagged" scores. Since the latter set of acceptable results include "OK-Tagged" SDUs, this accounts for the seemingly better performance on out-of-domain SDUs than on in-domain SDUs. For speech recognized input, we used the first-best hypotheses of the speech recognizer. The speech recognition word accuracy for this set was 66.8%.

The results shown in Figure 7 are from a large-scale evaluation of Spanish-to-English speech translation conducted in October 1995. The word accuracy for speech recognition was 63.0%. The evaluation set consisted of sixteen unseen dialogues that contained over 349 utterances with 1090 SDUs. We chose three versions of both the GLR and Phoenix modules that coincided with the development periods ending in November 1994, April 1995 and October 1995. At the time of the evaluation, the October 1995 version represented the end of the most recent development period. Figure 7 shows the percent of acceptable translations with out-of-domain detection ("Perfect", "OK" and "OK Tagged") for these time periods. Here we see a steady rise in acceptable scores over time. This rise is the result of development of the various knowledge sources of the translation components during these time periods.

Transcribed Input

	GLR*	Phoenix
November-94	72	72
April-95	82	77
October-95	85	83

Output of the Speech Recognizer

	GLR*	Phoenix
November-94	49	57
April-95	55	62
October-95	58	66

Fig. 7. Development of Spanish-to-English Translation Quality over Time

5 Summary and Conclusions

The evaluation of a speech translation system must provide a meaningful and accurate measure of its effectiveness. In order to accomplish this, it is essential that the evaluation be conducted on sets of "unseen" data that reflect translation performance under real user conditions. The evaluation procedure must neutralize subjectivity in scoring, take into account utterance length and complexity, compensate for data which is not relevant to the domain being evaluated, and employ a consistent set of criteria for judging translation quality.

Our end-to-end evaluation procedure described in this paper allows us to consistently and inexpensively measure the overall performance of our speech translation system. Our experience has shown that conducting frequent end-to-end evaluations is an effective tool for the development of our system and measuring its performance over time.

Acknowledgements

The work reported in this paper was funded in part by a grant from ATR - Interpreting Telecommunications Research Laboratories of Japan. We are grateful for ATR's continuing support of our project. We also thank Carol Van Ess-Dykema and the United States Department of Defense for their support of our work. We would like to thank all members of the JANUS teams at the University of Karlsruhe and Carnegie Mellon University for their dedicated work on our many evaluations.

References

1. R. Cole, J. Mariani, H. Uszkoreit, A. Zaenen and V. Zue editors. Chapter 13: Evaluation *Survey of the State of the Art in Human Language Technology*, Center for Spoken Language Understanding, Oregon Graduate Institute of Science and Technology, 1996. *http://www.cse.ogi.edu/CSLU/HLTsurvey/HLTsurvey.html.*

2. R. Hausser. Principles of Computational Morphology. Technical Report, Laboratory for Computational Linguistics, Carnegie Mellon University, Pittsburgh, PA, 1989.

3. A. Lavie. An Integrated Heuristic Scheme for Partial Parse Evaluation, *Proceedings of the 32nd Annual Meeting of the ACL (ACL-94), Las Cruces, New Mexico, June 1994.*

4. A. Lavie and M. Tomita. GLR* - An Efficient Noise Skipping Parsing Algorithm for Context Free Grammars, *Proceedings of the third International Workshop on Parsing Technologies (IWPT-93), Tiburg, The Netherlands, August 1993.*

5. L. Levin, D. Evans, and D. Gates. *The ALICE System: A Workbench for Learning and Using Language. Computer Assisted Language Instruction Consortium (CALICO) Journal*, Autumn 1991, 27–56.

6. L. Mayfield, M. Gavaldà, Y-H. Seo, B. Suhm, W. Ward, A. Waibel. "Parsing Real Input in JANUS: a Concept-Based Approach." In *Proceedings of TMI 95.*

7. C. P. Rosé, B. Di Eugenio, L. S. Levin, and C. Van Ess-Dykema. Discourse processing of dialogues with multiple threads. In *Proceedings of ACL'95, Boston, MA*, 1995.

8. B. Suhm, P. Geutner, T. Kemp, A. Lavie, L. J. Mayfield, A. E. McNair, I. Rogina, T. Sloboda, W. Ward, M. Woszczyna and A. Waibel. JANUS: Towards Multilingual Spoken Language Translation. In ARPA Workshop on Spoken Language Technology, 1995.

9. A. Waibel. Translation of Conversational Speech. Submitted to IEEE Computer, 1996.

10. A. Waibel, M. Finke, D. Gates, M. Gavalda, T. Kemp, A. Lavie, L. Levin, M. Maier, L. Mayfield, A. McNair, I. Rogina, K. Shima, T. Sloboda, M. Woszczyna, T. Zeppenfeld, P. Zhan. JANUS II: Advances in Spontaneous Speech Translation. Submitted to ICASSP, 1996.

11. M. Woszczyna, N. Aoki-Waibel, F. D. Buo, N. Coccaro, K. Horiguchi, T. Kemp, A. Lavie, A. McNair, T. Polzin, I. Rogina, C. P. Rosé, T. Schultz, B. Suhm, M. Tomita, and A. Waibel. JANUS-93: Towards Spontaneous Speech Translation. In *Proceedings of IEEE International Conference on Acoustics, Speech and Signal Processing (ICASSP'94)*, 1994.

A Task-Based Evaluation
of the TRAINS-95 Dialogue System*

Teresa Sikorski and James F. Allen

University of Rochester, Rochester, NY 14627, USA

Abstract. This paper describes a task-based evaluation methodology appropriate for dialogue systems such as the TRAINS-95 system, where a human and a computer interact and collaborate to solve a given problem. In task-based evaluations, techniques are measured in terms of their effect on task performance measures such as how long it takes to develop a solution using the system, and the quality of the final plan produced. We report recent experiment results which explore the effect of word recognition accuracy on task performance.

1 Introduction

TRAINS-95 is the first end-to-end implementation in a long-term effort to develop an intelligent planning assistant that is conversationally proficient in natural language. The initial domain is a train route planner, where a human manager and the system must cooperate to develop and execute plans [1]. TRAINS-95 provides a real-time multi-modal interface between the human and computer. In addition to making menu selections and clicking on objects using a mouse, the user is able to engage in an English-language dialogue with the computer using either keyboard or speech input.

The impetus behind the development of TRAINS-95 was the desire to implement and thereby test computational theories of planning, natural language, and dialogue, which have long been focal research areas of the Computer Science Department at the University of Rochester. Once the first version of TRAINS-95 was operational in January 1995, the question of how to evaluate the system needed to be addressed. Since this was just a first step in an incremental development, we wanted an evaluation method that would allow us to easily measure the progress made by subsequent versions of the system, and the effect of various design decisions on the performance of the system as a whole. Furthermore, we wanted to be able to use the evaluation results to guide us in system debugging, and to some extent focus our future research.

* Funding was gratefully received from NSF under Grant IRI-90-13160 and from ONR/DARPA under Grant N00014-92-J-1512. Many thanks to George Ferguson for developing the on-line tutorial, Eric Ringger for compiling the word recognition accuracy figures, Amon Seagull for advice on statistical measures, and Peter Heeman for numerous helpful comments. Thanks also to Mike Tanenhaus and Joy Hanna for their suggestions on the experimental design.

Standard accuracy models used to evaluate speech recognition and data base query tasks such as ATIS [4] are not appropriate for a dialogue evaluation. There is no right answer to an utterance in a dialogue. Rather, there are many different possible ways to answer, each of them equally valid. Some may be more efficient along some dimension or another, but there is no single best answer.

There is also a range of technology-based evaluations that could be performed, such as checking whether some agreed upon syntactic structure is produced, checking if referring expressions are correctly analyzed, or possibly checking if the right speech act interpretation is produced. These techniques are useful, but miss the real point. For instance, does it matter if the parse structures produced are faulty in some way if this is later compensated for by the discourse processing and the system responds in a way that best furthers the task the human is performing? The ultimate test of a dialogue system is whether it helps the user in performance of some task. This means that measures like time to completion and the effectiveness of the solution produced are critical. Of course, accuracy measures provide insight into various components of the system and thus will always be useful. But they will never answer the ultimate question about how well the system works. Accepting this argument, however, involves a shift of perspective from viewing the problem as a spoken language understanding problem to a problem that is more closely related to human factors and human computer interfaces.

The experiment described here is our first attempt to perform a task-based evaluation of a dialogue system. For our application, the task-based evaluation was performed in terms of two parameters: time to task completion and the quality of the solution. Our measure of solution quality was based on whether routes were planned to move trains from an initial configuration to a requested final configuration and, in cases where these goals were achieved, the number of time units required to travel to planned routes.

Our task-based evaluation methodology is domain-independent in that it can be applied to any system in which there are objectively identifiable goal states and solution quality criteria. For most planning applications, these are realistic constraints.

The solution quality criteria in the TRAINS-95 domain are extremely simple, but our method is extensible to future versions of the system where there will be an interplay between costs of various resources, all of which can be ultimately translated to a monetary cost. Once the goal state and solution quality criteria are defined in objective terms, the evaluation can be completely automated. This gives the task-based evaluation method a significant advantage over other techniques [7, 12] where human evaluators must intervene to examine individual responses and assess their correctness. This human intervention is both costly and introduces a subjective element that is unnecessary when task-based evaluation is applicable.

1.1 Evaluation Goals

By performing the experiment described herein, we have attempted to provide a quantitative evaluation of the level of success achieved by the TRAINS-95 implementation. A primary goal of TRAINS-95 was to develop a dialogue system that could exhibit robust behavior despite the presence of word recognition errors. The overall goal of the evaluation was to test this robustness. Other issues addressed in the evaluation include the following:

- the level of training required to use the system effectively
- establishment of an evaluation methodology and a baseline against which to evaluate future versions of the system
- identification of system deficiencies
- observations regarding user input mode preferences

1.2 Hypotheses

Our initial hypotheses were:

- Speech input is more time-efficient than keyboard input for accomplishing routing tasks using the TRAINS-95 system despite the presence of word recognition errors.
- Subjects with minimal training can accomplish routing tasks using the TRAINS-95 system.
- If word recognition accuracy is poor, the time taken to accomplish the task will increase.
- Users of the TRAINS-95 system prefer speech input over keyboard input.

Below we report the results of experiments conducted to test these hypotheses.

2 The System

The domain in TRAINS-95 is simple route planning. The user is given a map on a screen showing cities, connections, and the locations of a set of trains (see Figure 1), and a specification of a set of destination cities where trains are needed. The task is to plan routes to take the trains from the initial locations to the destinations. The route planner used by the system is deliberately weak so that interaction is needed to find good plans. Specifically, the planner cannot find routes longer than four hops without an intermediate city, and when it can generate a route, it randomly selects among the possibilities.

At the top of the TRAINS-95 software architecture are the I/O facilities. The speech recognition system is the SPHINX-II system from CMU[5]. The speech synthesizer is a commercial product: the TRUETALK system from Entropics. The rest of the system was built at Rochester. The display supports a communication language that allows other modules to control the contents of the display. It also handles keyboard input. The speech recognition output is passed through a statistical error-correcting post-processor developed at the University of Rochester[8]. The parser, accepts input either from the post-processor (for speech) or the display manager (for keyboard), and produces a set of speech act interpretations that are passed to the discourse manager. The discourse manager is itself composed of a range of subcomponents handling reference, speech act interpretation and planning (the verbal reasoner), and the back-end of the system: the problem solver and domain reasoner. When a speech act is planned for output, it is passed to the generator, which constructs a sentence and passes it to both the speech synthesizer and the display. The generator is a simple template-based system. It uses templates associated with different speech act forms that are instantiated with descriptions of the particular objects involved. The form of these descriptions is defined individually for each class of objects in the domain.

In order to stress the system in our robustness evaluation, we used the ATIS language model provided from CMU rather than a model based on TRAINS data. This system yields an overall word error rate of 30%, as opposed to a 20% error rate that we can currently obtain by using language models trained on our TRAINS corpus. While this accuracy rate is significantly lower than often reported in the literature, remember that most speech recognition results are reported for read speech, or for non natural-dialogue applications such as ATIS. Natural dialogue systems encourage a more spontaneous form of speech that is much more difficult to recognize.

To illustrate the typical complexity of spoken interactions, an example spoken dialogue from the experiment is given below. In the example scenario, the trains were initially at Charlotte, Atlanta and Cleveland and needed to be moved to Boston, Chicago and Montreal. In the transcript, we show the input as it was produced by the speech recognizer in ALL CAPS followed by what was actually said in italics. When the recognition is perfect, we just show the speech recognition output. The dialogue has a 82.2% word recognition accuracy.

SYS: OK, I think I'm ready to start.

USR: THE OKAY LET'S STEAK THE TRAIN FROM ATLANTA
AT TO CHICAGO VIA LEXINGTON IN INDIANAPOLIS
*Okay, let's take the train from Atlanta to Chicago via
Lexington and Inidanapolis.*

SYS: Yes

SYS: City Cincinnati is congested due to unusually heavy traffic.

SYS: City Columbus is congested due to unusually heavy traffic.

SYS: An additional 5 hours will be needed to travel through them.

USR: OKAY LET'S GO VIA INDIANAPOLIS

SYS: OK

USR: LIST STEAK THE TRAIN FROM CHARLOTTE TO BOSTON
Let's take the train from Charlotte to Boston.

SYS: I need help choosing a route from Charlotte to Boston.

USR: GO VIA RICHMOND AND PHILADELPHIA
SYS: Yeah
USR: OKAY LET'S GO FROM CLEVELAND TO MONTREAL
SYS: Yep
SYS: The terminal at City Buffalo is delaying traffic due to localized heavy winds.
SYS: Trains will take an additional 5 hours to move through there, due to decreased visibility.
USR: GO VIA DETROIT
SYS: Yeah
USR: AND THAT
   I'm done.
SYS: Excuse me?
USR: I'M DONE
SYS: Please confirm that you are done, using the dialogue box.

---

3   Experimental Design

In the experiment, input medium and word recognition accuracy were varied to determine their effect on our task performance measures (time to arrive at a solution and quality of the solution).

All subjects were given identical sets of 5 routing tasks to perform, in the same order. Half of the subjects were asked to use speech for the first task, keyboard for the second, speech for the third and keyboard for the fourth. The other half used keyboard for the first task and then alternated. All subjects were given a choice of whether to use speech or keyboard input to accomplish the final task.

While our evaluation appears similar to HCI experiments on whether speech or keyboard is a more effective interface in general (cf. [6]), this comparison was not actually our goal. Rather, we were using the various input media as a way of manipulating the input error rate. While keyboard performance is not perfect because of typographical errors (we had a 5% error rate on keyboard input, excluding self-corrections, During The evaluation), it is considerably less error prone than speech.

In addition to the differences in "word recognition accuracy" between keyboard and speech input, word recognition accuracy was further varied by using an error-correcting speech recognition post-processor for half of the subjects (four subjects using keyboard input first, four using speech input first). The speech recognition post-processor has been successful in improving the word recognition accuracy rate by an additional 5%, on average, in the TRAINS-95 domain.

The routing tasks were chosen with the following restrictions to ensure non-trivial dialogues and avoid drastic complexity variation among tasks:

- Each task entailed moving three trains to three cities, with no restriction on which train was destined for which city.

- In each scenario, three cities had conditions causing delays.
- One of the three routes in each scenario required more than four time units ("hops") to travel.

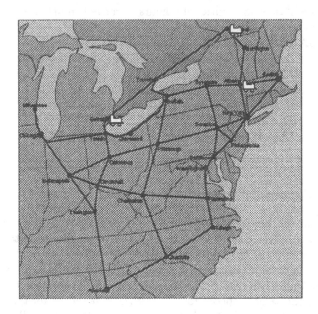

Fig. 1. TRAINS-95 Map

3.1 Experimental Environment

The experiment was performed over the course of a week in November 1995. Each of the sixteen subjects participated in a session with the TRAINS-95 system which lasted approximately 45 minutes.

All sixteen sessions were conducted in the URCS Speech Lab using identical hardware configurations. The software and hardware components used in the experiment included:

- A Sphinx-II speech recognizer developed at CMU, running on a DEC Alpha
- TRAINS-95 version 1.3 and the speech recognition post-processor running on a SPARCstation 10
- TrueTalk, a commercial off-the-shelf speech generator (available from Entropics, Inc.) running on a SPARCstation LX

Subjects, working at a Sun SPARCstation, wore a headset with a microphone to communicate with the speech recognizer. The TRAINS-95 system uses a click-and-hold protocol for speech input. The acoustic data from this speech input was recorded and later transcribed manually. Word recognition accuracy was computed by comparing these transcripts of what the subject actually said to transcripts of parser input from the speech recognition module and postprocessor.

The TRAINS-95 system communicated with the subjects verbally using the speech generator, visually by highlighting items on the map, and textually, through a text output window and dialogue boxes.

An example of the initial configuration of a railway map used by the TRAINS-95 system appears in Figure 1. The actual system uses color displays. As a route is planned, the proposed route is highlighted in a unique color. Cities that the system has understood to be goals or has identified to the user as causing delays are also highlighted.

Sixteen subjects for the experiment were recruited from undergraduate computer science courses. None of the subjects had ever used the TRAINS-95 system before, and only five reported having previously used *any* speech recognition system. All subjects were native speakers of American English.

3.2 Procedure

The four phases of the experiment are outlined below.

Tutorial The subject viewed an online tutorial lasting 2.4 minutes. The tutorial, which was developed for the purpose of the experiment, described how to interact with the TRAINS-95 system using speech and keyboard input, and demonstrated typical interactions using each of these media. Although the demonstration gave the subject an indication of how to speak to the system through the examples in the tutorial, the subject was given no explicit direction about what could or could not be said. The tutorial encouraged the subject to "speak naturally, as if to another person". While we are aware that human-computer dialogue is significantly different than human-human dialogue, there is evidence that such instructions given to test subjects does significantly reduce unnatural speech styles such as hyperarticulation [10].

Since it was important for the subject to understand how solution quality would be judged, the tutorial also explained how to calculate the amount of time a route takes to travel. For simplicity, subjects were told to assume that travel between any two cities displayed on the map takes one time unit. Additional time units were charged if the planned route included cities experiencing delays or if different train routes interfered with one another. (The system informs the user if either of these situations exist.)

Practice Session The subject was allowed to practice both speech and keyboard input before actually being given a task to accomplish. At the outset of the practice session, the subject was given a list of practice sentences that was prepared by a TRAINS-95 developer. Only the display and speech input modules were running during the practice session. Although there was no time limit, subjects spent no more than two minutes practicing, and several subjects chose not to practice keyboard input at all. During the practice session, a separate window displayed the output from the speech recognition module in textual format, indicating what was being "heard" by the system. This gave the subject an opportunity to experiment with different speech rates, enunciations, and microphone positions in order to accomodate the speech recognizer. The subject also learned to avoid common input errors such as beginning to speak before clicking, and releasing before completing an utterance.

Task Execution At the outset of each task, the subject was handed an index card with the task instructions and a map highlighting the destinations. The index cards specified which input medium was to be used, the destinations of the trains, and additional information about cities to be avoided. The subject didn't know the initial location of the trains until the map with the initial configuration was displayed on the monitor. The instructions for the first two tasks were simply to plan routes to get the trains to their destinations. Instructions for the three remaining tasks asked the subjects to find efficient routes as quickly as possible. During this phase of the experiment, the subject had no interaction with the experimenter and the speech recognition module's output was not displayed.

Questionnaire After completing the final task, the subject was given a questionnaire that solicited the impressions about the cause of any difficulty encountered, reasons for selecting speech or keyboard input for the final task, and recommendations for improvements.

4 Experiment Results

Results of the experiment relevant to our hypotheses are as follows:

- Of the 80 tasks attempted, there were 7 tasks in which the stated goals were not met. (Note that the figures do not include dialogues in which the goals were not accomplished.)
- Of the 16 subjects, 12 selected speech as the input mode for the final task and 4 selected keyboard input.
- Figures 2 and 3 show that the plans generated when speech input was used are of similar quality to those generated when keyboard input was used. However, for each task, the amount of time to develop the plan was significantly lower when speech input was used. Speech input was 28 − 49% faster. This performance of speech input over keyboard input in our experiment is in contrast with experimental results obtained on some previous systems [9]. The experiment results are consistent with our expectations that as speech

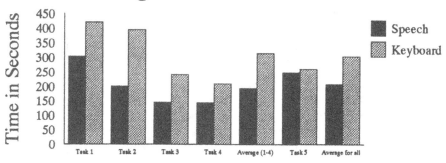

Fig. 2.

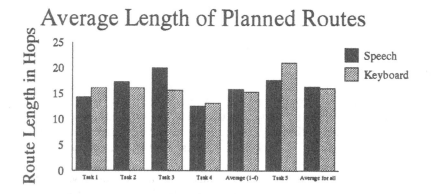

Fig. 3.

recognition technology improves and robust techniques for dialogue systems are developed, human-computer interaction using speech input will become increasingly more efficient in comparison with textual input as is the case with human-human interaction [3].

- Figure 4 plots word recognition accuracy versus the time to complete the task for each of the 5 tasks.

Note that in Figures 2 and 3, we give the average for Tasks 1-4, as well as the average for all 5 tasks. The average for Tasks 1-4 may be a more accurate point of comparison since in Task 5 the subjects were using the medium with which they were most comfortable.

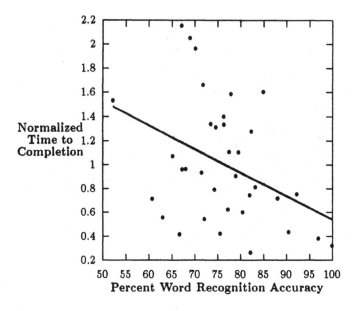

Fig. 4.

5   Discussion

In each task where the goals were not met, the subject was using speech input. Interestingly, there was no particular task that was troublesome and no particular subject that had difficulty. Seven different subjects had a task where the goals were not met, and each of the five tasks was left unaccomplished at least once.

A review of the transcripts for the unsuccessful attempts revealed that in three cases, the subject misinterpreted the system's actions, and ended the dialogue believing that all the goals had been met. In each of these cases, the misunderstanding occured when the system identified a destination mentioned by the subject by highlighting the city on the map. The subjects took that action as meaning that a route had been planned between the initial location of the train and the destination even though no proposed route was highlighted.

Each of the other four unsuccessful attempts resulted from a common sequence of events: after the system proposed an inefficient route, word recognition errors caused the system to misinterpret rejection of the proposed route as acceptance. The subsequent subdialogues intended to improve the route were interpreted to be extensions to the route, causing the route to "overshoot" the intended destination. In each of these scenarios, the subject did get the train to pass through the destination specified in the task instructions, but the planned route did not terminate at the required city.

One of our hypotheses was that as word recognition accuracy degraded, the time to arrive at a solution would increase. Figure 4 depicts the relationship between task completion time and word recognition accuracy observed during the experiment, where the time to completion figures have been normalized by task. We expected to find a strong negative correlation in Figure 4. However, the correlation coefficient of the best fit line in the figure is only 15.7%. We've identified three possible explanations for why we didn't find as significant a correlation between word recognition accuracy and time to completion as we expected:

Robust Parsing The word recognition accuracy measurement used in the evaluation was computed as follows:

$$WRA = \frac{N - D - S - I}{N}$$

where
WRA = word recognition accuracy
N = the number of words actually spoken,
D = the number of words deleted,
S = the number of word-for-word substitutions, and
I = the number of words inserted

The TRAINS-95 system makes use of a robust chart parser which is often able to form an interpretation for an utterance even when many of the words are misrecognized. This makes it less important how many words are correctly recognized as *which* words are correctly recognized. Since the word recognition accuracy measure we used treated all words as having equal significance, it was not as telling a statistic as initially expected.

Nonunderstanding vs. Misunderstanding Many times, the TRAINS-95 system cannot form an interpretation of an utterance due to poor word recognition. When the TRAINS-95 system experiences nonunderstanding, it enters into a clarification subdialogue, and takes no action based on the misunderstood utterance. The user is then able to repeat or rephrase the utterance, and the dialogue continues. When nonunderstanding occurs, both the user and the system are aware of the situation and are able to quickly and easily rectify the situation.

Misunderstandings, on the other hand are often undetected initially, and are more time-consuming to fix, since the system *has* taken action based on its erroneous interpretation of the utterance. Misunderstandings are therefore significantly more detrimental to task performance than instances of nonunderstanding. Since misunderstandings occur even in dialogues where the subject experiences relatively good speech recognition, high word recognition accuracy and low time to completion do not always correlate.

Our experiment indicates that a robust approach (i.e. one that tries to form an interpretation even when there's low confidence in the input) can create a high variance in the effectiveness of an interaction. Note, however that a comparison of two dialogue systems - one taking a conservative approach that only answers when it is confident (cf. [11]), and a robust system that proceeds based on partial understanding - showed that the robust system was significantly more successful in completing the same task [7].

Random Routes Designers of the TRAINS-95 system feared that since the initial domain is so simple, there would be very limited dialogue between the human and the computer. In order to stimulate dialogue, the designers used a deliberately weak planner that has the following properties:

- The planner needs to ask the human for help if given a destination more than four "hops" from the original location of the train.
- The planner randomly selects between possible routes when the human does not specify a specific route.

The second property entered an amount of nondeterminism into the experiment that couldn't be compensated for by the small number of subjects we used. Some subjects that had good word recognition accuracy had to spend a significant amount of time improving inefficient routes generated by the planner. In some other tasks, where poor recognition accuracy was a problem, the planner fortuitously generated efficient routes the first time.

Exacerbating the problem was the system's poor handling of subdialogues where the subject attempted to modify routes. Most of the subjects expressed their frustration with this aspect of the system on the questionnaire. After several interactions with the system aimed at improving routes, subjects many times either gave up on the task or restarted the scenario, losing good routes that had been previously completed. Other subjects left the inefficient routes as they were, adversely affecting the quality of the solution. These problems reveal a need for better handling of corrections, especially as resumptions of previous topics.

6 Future Directions

Development of TRAINS-96 is near completion. TRAINS-96 will have an expanded domain involving realistic distances and travel times between cities with associated costs. Time and cost contraints will be added to the given tasks. Since richer dialogues will be a natural consequence of the richer domain, the system will no longer contain a weak planner, thus eliminating many of the problems artificially introduced during the evaluation of TRAINS-95.

A fundamental design decision in TRAINS-95 was to choose a specific interpretation when faced with ambiguity, thus risking misinterpretation, rather than entering into a clarification subdialogue. We plan to evaluate the effectiveness of this strategy in later versions of the system.

Future evaluations will also involve a comparison of the effectiveness of a human solving a given routing task alone (with the aid of a spreadsheet and calculator) versus a human solving the same task with the assistance of the TRAINS system.

Future evaluations may also include additional independent variables such as task complexity and additional interface modalities. We have hypothesized that the expanded domain will stimulate more spontaneous dialogue, and we would therefore like to perform experiments that evaluate the richness and spontaneity of the language used by the subjects using new versions of the system. This type of evaluation will require that we establish a fairly objective measure of spontaneity and richness, perhaps based on the extent of the vocabulary used by subjects, the length of subjects' utterances, and the presence of various linguistic phenomena such as speech repairs, anaphoric reference, etc.

In our evaluation of the TRAINS-95 system, we relied on word recognition accuracy to give us an indication of the system's level of understanding of the dialogue. As Figure 4 demonstrates, word recognition accuracy did not have an especially strong correlation with task performance measured in terms of time to solution in our experiment. In the TRAINS-96 evaluation, we intend to employ more sophisticated techniques that will measure understanding of utterances and concepts rather than words, perhaps based on the recognition accuracy of certain key words used by the robust chart parser. Recent research indicates a linear relationship between word recognition accuracy and the understanding of utterances [2].

The evaluation of the TRAINS-95 system had coarse granularity in that it measured the amount of time taken to complete the task and the quality of the solution as a whole. During the TRAINS-96 evaluation we hope to also be able to perform an evaluation of the system's performance during various types of subdialogues. Unfortunately, an *automatic* task-based analysis will not always be possible at a subdialogue level, but, in general, a transcript review will indicate what the goal of the subdialogue was and whether the goal was met. Evaluating at a subdialogue granularity may give system developers a better indication of where system improvements are most needed.

References

1. J. F. Allen, G. Ferguson, B. Miller, and E. Ringger. Spoken Dialogue and Interactive Planning. In *Proceedings of the ARPA SLST Workshop*, San Mateo California, January 1995. Morgan Kaufmann.
2. M. Boros, W. Eckert, F. Gallwitz, G. Görz, G. Hanrieder, H. Niemann. Towards Understanding Spontaneous Speech: Word Accuracy Vs. Concept Accuracy. In *Proceedings of the International Conference on Spoken Language Processing*, Philadelphia, Pennsylvania, October 1996.
3. P. Cohen and S. Oviatt. The Role of Voice Input for Human-Machine Communication. In *Proceedings of the National Academy of Sciences*, 1994.
4. L. Hirschman, M. Bates, D. Dahl, W. Fisher, J. Garofolo, D. Pallet, K. Hunicke-Smith, P. Price, A. Rudnicky and E. Tzoukermann. Multi-Site Data Collection and

Evaluation in Spoken Language Understanding. In *Proceedings of the ARPA Human Language Technology Workshop*, Princeton, New Jersey, March 1993. Morgan Kaufmann.

5. X. D. Huang, F. Alleva, H.W. Hon, M. Y. Hwang, K. F. Lee, and R. Rosenfeld. The Sphinx-II Speech Recognition System: An Overview. *Computer, Speech and Language*, 1993.

6. S. Oviatt and P. Cohen. The Contributing Influence of Speech and Interaction on Human Discourse Patterns. In J. W. Sullivan and S. W. Tyler (eds), *Intelligent User Interfaces*. New York, New York. 1991. Addison-Wesley.

7. J. Polifroni, L. Hirschman, S. Seneff, and V. Zue. Experiments in Evaluating Interactive Spoken Language Systems. In *Proceedings of the DARPA Speech and Natural Language Workshop*, Harriman, New York, February 1992. Morgan Kaufmann.

8. E. Ringger and J. F. Allen. Error Correction Via A Post-Processor For Continuous Speech Recognition. *Proceedings of ICASSP-96*, Atlanta Georgia, May 1996.

9. A. Rudnicky. Mode Preferences in a Simple Data Retrieval Task. In *Proceedings of the ARPA Human Language Technology Workshop*, Princeton, New Jersey, March 1993. Morgan Kaufmann.

10. E. Shriberg, E. Wade, and P. Price. Human-Machine Problem Solving Using Spoken Language Systems (SLS): Factors Affecting Performance and User Satisfaction. In *Proceedings of the DARPA Speech and Natural Language Workshop*, Harriman, New York, February 1992. Morgan Kaufmann.

11. R. Smith and R. D. Hipp. *Spoken Natural Language Dialog Systems: A Practical Approach*, Oxford University Press. 1994.

12. S. Walter. Neal-Montgomery NLP System Evaluation Methodology. In *Proceedings of the DARPA Speech and Natural Language Workshop*, Harriman, New York, February 1992. Morgan Kaufmann.

Springer
and the
environment

At Springer we firmly believe that an
international science publisher has a
special obligation to the environment,
and our corporate policies consistently
reflect this conviction.

We also expect our business partners –
paper mills, printers, packaging
manufacturers, etc. – to commit
themselves to using materials and
production processes that do not harm
the environment. The paper in this
book is made from low- or no-chlorine
pulp and is acid free, in conformance
with international standards for paper
permanency.

Springer

Lecture Notes in Artificial Intelligence (LNAI)

Lecture Notes in Computer Science